Praise

CW00346934

'I highly recommend *She Chose Tech* to anyone interested in broadening their knowledge in the tech sector. This resource covers relevant and up-to-date topics clearly and concisely, drawing on Sonal's personal experience in the field. Her inspiring story and the accounts of women she has interviewed make the book engaging and informative. Whether you're considering a career in tech or seeking inspiration in the early or mid-level stages of your journey, this book is an excellent choice.'

— **Hira Ali**, Founder of Advancing Your Potential and Author of *Her Way To The Top: A guide to smashing the glass ceiling* and *Her Allies: A practical toolkit to help men lead through advocacy*

'An easy to digest introduction to technology careers through the eyes of an experienced technologist. If you are looking to break into the industry, this book will provide food for thought on the differing roles in the industry and emerging technology. Well done, Sonal, for inspiring others to pursue their careers in tech through your own story.'

— **Vanessa Vallely OBE**, CEO, WeAreTheCity and WeAreTechWomen

'In this book, Sonal Shah unveils a compelling roadmap for women to conquer the tech industry while demystifying its complexities. Sonal empowers us to embrace technology fearlessly, fostering a brighter, more inclusive future. A must-read for aspiring women in tech! This book is suitable for anyone wishing to learn about technology.'

— **Bianca Miller-Cole**, Entrepreneur and Personal Branding Expert

'Sonal has written an easy-to-read book for anyone wishing to learn. *She Chose Tech* covers some interesting and compelling points, all put across simply so anyone can understand what sometimes can be made complex in technology.

'This book will benefit many readers, and Sonal provides value by sharing her own experience and discussing the latest emerging technologies, such as Artificial Intelligence. She outlines why we need more diversity in technology and, importantly, de-mystifies reasons for all future technology to be built for successful use by all.

'Interviews with professional women working in technology, including Dr Anne-Marie Imafidon MBE, further light up this inspiring read, helping those interested to learn become more aware of tech or even consider joining the industry to better themselves.

'There is something here for all levels; an informative read for anyone wishing to take action to obtain the change they desire.'

— **Dr Byron Cole**, HonDBA, Multi Award Winning Entrepreneur, *Sunday Times* Best-Selling author

'*She Chose Tech*'s Sonal Shah builds on her amazing work as a diversity and technology advocate with a truly practical guide to entering and navigating the world of tech. Sonal's authentic and jargon-free approach demystifies this sector and helps the reader understand and access the wealth of career opportunities available in this ever-changing world. Sonal's dedication to advocating for and teaching others alongside building her own career is an inspiration to us all.'

— **Emilie West**, Founder of Alchemy Coaching and Author of *Maximum You: Harnessing the power of authenticity*

'Sonal takes her remarkable journey as a diversity and technology advocate to the next level, and as a guiding light for anyone seeking to know more or thrive in technology. Her approach is refreshingly authentic, where she shares her knowledge and experience, which is nothing short of inspiring, simply helping anyone to know more or helping make the tech sector accessible to all.'

— **Adam Stott**, Award-Winning Entrepreneur

'*She Chose Tech* is an enlightening reading experience. Through Sonal's ability to demystify some of the intricacies, she covers emerging technologies, such as AI, crypto and blockchain, together with other skills such as networking and personal branding.

'Sonal's book is powerful in this era of limitless potential where we can all pick tech and pave our extraordinary ways. This book is a testament to her commitment and for anyone seeking to know more or thrive in technology.'

— **Helen Disney**, Founder of Unblocked, Former Editorial Writer at *The Times*, Director of Blockchain and Digital Assets at The Realization Group

'We still have too few women working in tech, and often women have a variety of reasons for thinking the tech sector or tech jobs are not for them. I was one of them until I got involved in cryptocurrency and blockchain and realised these technologies needed my communication and storytelling skills, even though I never saw myself as a "techie". Sonal Shah is doing important work encouraging more women into the tech world by sharing her own story and those of others so that the next generation will be inspired to see tech for the rewarding career and opportunity it represents.'

— **Liam Ryan**, Property and Business Entrepreneur

She Chose Tech

SONAL SHAH

The essential guide to inspire and empower women in tech

First published in Great Britain in 2023
by Rethink Press (www.rethinkpress.com)

© Copyright Sonal Shah

All rights reserved. No part of this publication may be reproduced, stored in or introduced into a retrieval system, or transmitted, in any form, or by any means (electronic, mechanical, photocopying, recording or otherwise) without the prior written permission of the publisher.

The right of Sonal Shah to be identified as the author of this work has been asserted by her in accordance with the Copyright, Designs and Patents Act 1988.

This book is sold subject to the condition that it shall not, by way of trade or otherwise, be lent, resold, hired out, or otherwise circulated without the publisher's prior consent in any form of binding or cover other than that in which it is published and without a similar condition including this condition being imposed on the subsequent purchaser.

978-1-78452-602-3

I dedicate this book to my dear late father and late mother. They exhibited a strong work ethic, honesty and strength to overcome barriers. A constant source of love, encouragement and support, they were true role models who gave me everything I need.

I particularly express my gratitude to my brother and his family, whose support has been a constant source of strength.

This book is for anyone who is striving to achieve more and be more.

Contents

Foreword

With low numbers of women in the tech industry, *She Chose Tech* is a must read for anyone striving to make tech their career. It offers valuable insights and inspiration for breaking gender barriers in this dynamic industry.

I met Sonal some years ago through the LMF Network (Like Minded Females Network) and have witnessed her commitment to diversity and driving change in the technology industry, evident through her many awards and recognition. Sonal inspires us with her story of how she changed careers, openly sharing how others can repeat her successes. In an easy-to-understand way, she demystifies tech and uses her knowledge, journey and views about tech to empower and help women to achieve their goals.

Women are skilled, talented and ready; they merely need to be provided with guidance otherwise unavailable, and connections with mentors to open doors. As women of colour, or any women, including some from disadvantaged backgrounds and forgotten histories, we need mentors to help us understand how to ask for increased pay, take up space and remind us how important we are.

She Chose Tech presents great information, dispelling myths in tech, showing how to change careers with details about some common roles and skills required, and shares interesting points about personal branding and imposter syndrome. We also learn about emerging technologies, such as artificial intelligence, crypto and blockchain.

Through this book there are interviews with women, including Dr Anne-Marie Imafidon MBE, who are technology professionals and experts ready to share their experiences in tech careers.

My journey in technology since 2015 has had mentors and friends push me to strive harder, reminding me that the difference in diversity is me, the individual, and that to create inclusion, we need to take up the space rightfully ours. Sonal manages to pull that experience into her book – giving readers the confidence to excel in the industry.

Knowing the power of technology is important, but the industry, and the world, needs professionals who are representative of the many differences and data points that make us unique. Some say AI will take over the world – many fear it, but tech is not to be feared. More women are essential in the industry and this book illustrates the different ways to join, along with the many career opportunities available. Bias is covered and the need for diversity of thought in making the technology – which is important to ensure tech is made in the right way for all to use.

We must make sure the data we are feeding the algorithms is diverse, and that comes from welcoming people from all sections of our society into the tech industry to learn the craft.

She Chose Tech is a guide that we need, and we must pass it down to the generations climbing up the ladder to teach them what they don't know and remind them that they are tech's greatest asset.

Sonya Barlow
Founder of LMF Network, Award-Winning Entrepreneur, BBC Broadcaster, Diversity Coach, Travel and Tech Writer, MC and Author of *Unprepared to Entrepreneur*

Introduction

Are you a young graduate curious about a tech career? You may be early in your career and still deciding which direction to go. Perhaps you are looking for a change and wish to work for an employer who will appreciate your skills.

Tech is a field full of breakthroughs, creativity and research, where innovation solves the world's most pressing problems. A lot of purposeful work is done throughout the tech sector, and it is great to be part of the solutions. Wherever you are on your journey – from a student or graduate to a working professional, early to mid-career – or whether you are someone who impacts others, such as a coach, teacher, careers advisor, human resources (HR) professional or mentor, if you are interested in tech and want more knowledge and awareness on the subject, then this book is for you. Parents, you also play a crucial role in increasing awareness of women in tech. This book can assist you in supporting and influencing those you care about to join this dynamic industry – which will also increase the pipeline for women in this industry.

If you are curious and wish to learn more about this subject, *She Chose Tech* will show you what is possible and help you make a decision and take the next step. It will demystify tech, inspire you and show you what it is all about! You can use this book as a pick and learn to give you awareness and educate yourself.

I am shining a light on this sector because there are far fewer women than men in the tech profession. The 'Women in Tech' survey report 2023 states only 26% of tech professionals are women. If you are considering your future, or someone you know is, whatever your position and experience, tech offers a breadth of opportunities. Throughout the book, I will advise you on the potential of women in this sector and how to attract diversity into the world of tech.

As an ambitious woman myself, I am committed to championing diversity. Having won awards for my passion in addition to and outside the day job, I am well placed to explore and share my real-life experience with you in a straightforward manner – no need for Google! I have done this through sharing what I have learned from my journey to give you a head start, this book will teach you things you do not yet know, inspire you to explore things you want to know and encourage you to become curious.

Everyone can benefit from becoming more 'in the know' about tech careers and be more tech savvy, as it is not going away. If you are looking to find out more about tech, perhaps wondering how to join this industry, build your confidence and awareness or find a suitable role, I want to demystify it for you. If you want to know what crypto, blockchain or artificial intelligence (AI) is, I will share some insights into these areas and others for you.

As this is a vast subject, I have chosen some of the most relevant and topical subjects to enhance your understanding today. Depending on which area most interests you, once you have finished the book, you will be ready to focus and dive deeper into that area, researching it further. I have added references and further resources at the end of the book.

Things change and move fast in the tech industry, so I will give you a flavour of the way things are right now with examples across the chapters to whet your appetite. I will describe some of the most exciting women in tech and their routes into the industry, along with the common roles you will find.

I have been fortunate to be recognised with multiple industry-wide awards, such as WeAreTech100 and a

Rising Stars Award in Diversity from WeAreTheCity in 2018. I was also named one of the Most Influential Women in *Computer Weekly* in 2022 and 2023 and *Brummell Magazine* Inspirational Women 2023, but for me, the most exciting thing is to reach out to and talk with as many women as possible. I want to share the excitement of women in tech and emphasise that the door is open for anyone to be a part of this dynamic industry. I have enjoyed participating in various activities and being invited to speak within communities, professional settings and conferences to get my message across and share valuable insights. This book is simply one more avenue to share my knowledge and experience with a broader audience.

Through my experience, I have learned about personal branding approaches and networking, and I will teach you how to apply these skills to achieve your goals when working within this ever-changing sector. I have also fallen victim to perfectionism and imposter syndrome, common afflictions of successful people. We will discuss these in later chapters, so you can be aware of them and ready to circumvent these obstacles and apply these skills to achieve your goals.

One person alone cannot do everything in a complex environment; it takes many with a range of skills to develop products to meet all potential customers' needs in the right way. The more diverse the range of people combining their skills and experience, the more quickly better solutions are achieved to meet the fast-changing markets. Increasing the numbers of women in tech and showing our worth will help the industry gain more. The tech sector is starting to understand that diversity aids in achieving better outputs. We can all help make that a lasting change.

From a young age, girls need to be enticed into tech careers, even if they don't particularly enjoy science or technical problem-solving at school. Their experiences may be more favourable in areas other than the science, technology, engineering and maths (STEM) subjects, but their skills are still needed in tech. Non-technical opportunities abound within the industry, from marketing to managerial and you can indeed become a future tech leader.

There is no one path to join the tech sector from school, college or university, and no one path to change your career and succeed. If this is what you want, there are many ways into the tech industry, and having a journey that's different to your peers is okay. My journey and observations of others, and learning from them, have shown me what has been most helpful, so I want to make your path easier. Whether you are looking for starter positions, at a career crossroad or to pivot mid-career, I encourage you to join this dynamic industry; the opportunities are immense.

Many factors affect businesses, both globally and locally. At the time of writing, these include restarting post-Covid, and economic and environmental challenges. Despite the difficulties, your skills are needed in tech right now, and you will not be alone. There are women already in the industry who can guide you – we all have a role to play in helping others to thrive and supporting the change for a diverse workforce. If you want to achieve your goals and follow your dreams, I will share what I know to help you make that happen. This book will enable you to be genuinely 'with it' in the digital age, whether you're helping the retired or are indeed retired, having intelligent conversations with friends or colleagues, or a parent wanting to see your children or grandchildren make informed decisions.

Through the medium of this book, I will spread the word about the breadth of what is available to you. All thoughts, ideas or opinions expressed here are my own, and not a reflection on any of my former or current employers.

Nothing is easy, but tech can be richly rewarding for your mind, your finances and society as a whole. No matter why you choose to enter the tech industry, you have everything it takes to be successful. Get inspired and learn more about what interests you so you can be a trailblazer, a go-getter woman who excels in her role and, more importantly, loves it.

ONE

Techtopia: Your Journey Into The Electrifying World Of Technology

> 'The passion for stretching yourself and sticking to it, even (or especially) when it's not going well, is the hallmark of the growth mindset.'
> —Carol S Dweck

Carol Dweck is an American psychologist and one of the world's leading researchers on motivation and mindset. She is well known for her writings on the subject of growth mindset, and this quote from her excellent book *Mindset: Changing the way you think to fulfil your potential* is extremely relevant to the tech sector. Technology is often all about continual learning.

In this chapter, I will be sharing a little of my story and background, showing how I changed careers and started in tech. I will introduce the basics of getting into tech, and include statistics about women in this sector, along with some of the reasons why there are fewer females than males in the industry.

I do wish there were no exclusive efforts for the need to have particular actions such as having women in tech conferences, tech talks, meet-ups for women in tech, and many different community events. However, there is a need for more skilled tech workers in the UK and worldwide. With the demand for tech products and services, the lack of diversity in the industry is something to be mindful of, especially the lack of gender diversity, not only when we think of the skills gap.

The Tech Talent Charter – Diversity in Tech report 2023 shares: '...gender diversity decreases by 5% for senior tech roles, while ethnic diversity decreases by 12%.' This is not great for the tech industry. Without balanced gender input and diversity in general, we will not see the progress we require. It means that many developments, advancements and innovations will lack the more inclusive viewpoint that women can put forward.

For years, we have seen reports about the gender pay gap, which the Office for National Statistics (ONS) census 2021 tells us is calculated as the difference between the hourly earnings of men and women as a proportion of men's average hourly earnings (excluding overtime). The measure considers all jobs in the UK, but does not look at the differences in the salary of men and women in the same position, which is interesting.

The United Nations (UN) records state, 'Across the world, women still get paid 23% less than men. At the current rate of progress, no equal pay until 2069.' (UN Women.org, 2022.) I would love to see gender equality in pay scales, but these figures highlight the differences generally for women.

Let me share with you a personal story. Some years ago, I became bored with my career. I had worked my way up to manager level and saw the potential and excitement of the dynamic tech industry, not to mention the higher-earning possibilities and better prospects, so I decided to take a year out and complete a Master of Science (MSc) in computing and information technology (IT), as I knew nothing about it. I emphasise that a degree is not the only way into tech; it's simply what I personally chose to do. There are, and will continue to be, many more routes to enter the tech industry if you too want to change fields.

If you are in a management or leadership position, I urge you to recognise and grow the talent of the younger generations and help those in mid-career. We require role models in place for young women to see and recognise as they climb the ladder in their careers. If they cannot see it, they cannot be it, but as we have already discovered, the Tech Talent Charter 2023 tells us that gender diversity decreases by 5% for senior roles, and currently, only 26% – yes, 26%! – of the tech workforce is made up of women. We still have work to do! If you're a parent, you are ideally placed to guide your children, in particular your daughters, into this dynamic and exciting industry.

Tech bias

As I write this, I am thinking about how to improve the gender pay gap figures in general, and particularly in the tech industry. Diversity is essential to develop technology further and in the right way, with both male and female perspectives.

Let's take as an example the development of the original car seatbelt, which failed. Initially, the developers only considered tall, middle-aged males, not smaller people and definitely not petite women, so it was unsafe for many. Thank you to Caroline Criado Perez, who highlighted this and many examples in her book *Invisible Women Exposing Data Bias in a World Designed for Men*, showing where advancements and innovations could have been better if they hadn't been predominantly developed by men with men in mind. Did you know that until 2011, all car crash-test dummies were based on standard male figures? In her book, Criado Perez says, 'I can't be the only woman who ends up holding on to the seatbelt to avoid it strangling me.' This is one example that clearly shows the requirement for more diversity of thought to go into future developments, in the tech world and beyond.

If tech developers don't embrace diversity of thought, we will always have problems. In the early days of face recognition technology, certain people of colour who identified as Black or Asian were falsely identified 10 to 100 times more often than others. The face recognition technology was biased as the developers did not have enough data to represent the diverse range of people.

This topic is discussed further in a 2022 article by Bernard Marr which highlights how human beings choose the data that algorithms use in AI. They decide how the results of the algorithm are applied which can have biases because they are only as good as the data input by humans. The article goes on to say that '...81% of business leaders want government regulation to define and prevent AI bias'. This would reduce poor outcomes for certain people, requiring accountability and transparency from the producers of this technology.

In my interview with Dr Anne-Marie Imafidon MBE, which you will find in Chapter 3, she speaks about the collection of datasets used by AI, which is worthy of note as it relates to bias.

How I changed career

When I started out in tech years ago, I had no mentor or coach, so I did plenty of research before taking my leap of faith from a secure job into a complete career change. This was a particularly bold move, given that I knew nothing about computers or technology in general. I only knew about Microsoft Office, and I disliked science at school with a vengeance!

My point is that it can be done. You *can* make the move into tech whatever your background, and don't believe those who tell you that you cannot, because you can.

Stay with me to find out more, to realise that tech is not rocket science (pardon the pun here, you know what I mean!). Seriously, there is such a vast array of roles that tech touches every part of our lives. It can pay well, and there is so much demand for talent, you will always have work. Furthermore, it is not all about coding. May you become inspired and passionate about this field and find an area within tech you enjoy.

At school and college, I never felt that tech was for me. I thought it would be dull and full of science, and unfortunately, no one pointed out the benefits. Plus, there were few industry role models, which added to my lack of interest from a young age.

However, I have always loved writing and have often had work published, ranging from blogs to comments in well-known magazines to being asked for media comments on diversity, equity and inclusion (DE&I). Also, I love public speaking, where I can share messages at forums, conferences, workshops, schools and universities, or business talks with immediate audience feedback and interaction. You and I may meet in person at some stage, or you may already have heard me speak.

All these skills, and many more, are valuable in the tech industry. My aim with this book is to steer you in the right direction, giving you ideas so you can find exactly where you may fit. This applies if you are a student at school, college, university before you join the working world or even if you are already working.

First, however, you need to know what tech is and why you should consider it as a career.

What is going on for women in tech?

Technology is a progressive field, but one reason for the small number of women in the sector is the lack of role models. When females choose to study computer science degrees or tech courses, they tend to be outnumbered by males, which brings the challenge of being in the minority. It is the same at work. Some workplaces still have not cultivated diverse and inclusive environments, and although recently we have seen some changes, if an industry is heavily male-dominated, there is potentially a high rate of females not staying. They may not see a progression path because there are no female role models in their company to aspire to. All these reasons add to the sad statistic that only 26% of tech professionals are women.

Young women are often not encouraged to pursue STEM subjects early enough in education. As a female, you may have experienced this at school and college, and relate to my words. Those interested or just starting in tech may need access to opportunities and academic resources to help develop their skills.

For older readers who are parents or grandparents, please note it is essential for girls to be encouraged to take part in STEM subjects and activities and be encouraged at home and in the school environment. Please do think about this when you speak to the young in your family. Many free resources are available, shared later in the book, and there are communities to learn from if you are unsure what to say. You will also gain insights throughout this book.

Do not worry, though. Things are steadily progressing as the demand for skills is vital in this industry, and remember, you can become anything you put your mind to. Let's keep an open mind and be focused. If there is one thing tech does well, it's encouraging us to learn new things and stay updated continually. You will be fine if you choose this path.

You need to know which technology careers and roles are available and become aware of the types of skills you will need. However, things move rapidly in tech, with new positions constantly evolving. Even if you are in the minority at the moment, you can achieve anything if you put your mind to it. Do not be fearful. Things *will* change in tech, so please step out of your comfort zone and keep an open mind. By reading this book, you have already made a start.

The Department for Culture, Media, Digital and Sport estimates the digital skills gap costs the UK as much as £63 billion a year (Philp, 2022). To close this gap, it is vital for us all to know how we can contribute.

Every year, activists, advocates, experts and governments worldwide come together for the annual meeting of the UN 67th Commission for the Status of Women (CSW). In 2023, the commission convened under the theme: 'Innovation and technological change, and education in the digital age for achieving gender equality and the empowerment of all women and girls.' This adds weight to the importance of equalising the gender gap in technology.

The first commission dates back to 1947, just two years after the founding of the UN, when a group of fifteen women representing governments from around the world met in New York to begin building the international legal foundations of gender equality. These trailblazing female leaders took on projects to raise global awareness of women's issues and change legislation. For instance, the commission contributed to drafting the Universal Declaration of Human Rights and negotiated more gender-inclusive language. It drew up early international conventions on women's political rights, women's rights in marriage, equal pay for equal work and more.

You may wonder why I am even mentioning this. It shows the importance placed globally on gender equality as many issues arise from organisations not having a good balance of males and females. Tech is forever growing and is in demand worldwide, so it is on the minds of many key figures and influential people.

Why should women care about tech?

Of all the career fields available to us, tech is the vastest as it touches every area of our lives. It has many options and choices – for example, social media to connect people – as well as being creative and fun – for example, the innovative design of websites. Why should you, as a girl or woman, or even a girl or woman who is also in an ethnic minority group, care about that?

We need to increase the percentage of females in tech from 26%, which includes *only* under 4% who are women of colour. There are many opportunities that await you, should you choose to pursue a career in technology.

'You do not know what your abilities are until you make a full commitment to developing them,' says expert psychologist Carol S Dweck in her book *Mindset: Changing the way you think to fulfil your potential*. Worldwide, the tech industry currently has more jobs than people to fill them, so it is an excellent place to find well-paid careers to develop your abilities in a sector that needs more women. It is an exciting, changing place to find yourself in, where you keep up with dynamic and complex technology. Aside from the innovative opportunities in this growing industry, there is an excellent earning potential and chance of progression. This is your time to join this thriving and lucrative industry.

What is my 'why'? What inspired me to leave an unrelated job and take a year out to study tech so I could totally change my career years ago? What has made me stay?

For me, tech was and still is one of the few career fields that stands out. It is dynamic and fully utilises my interests, skills and passions while providing opportunity and earning power. I do not work in a full technical role; I work on various projects which give such great variety and, as mentioned before, tech touches all different parts in your work. Although

I have faced challenges, and not always of the good kind, I am so happy that I changed careers. I will share some of the challenges I faced from the perspective of being female in tech, and the lessons I learned, throughout the book to help you on your journey.

I love learning new things at work in the various projects I do and found my way into the project world. No two days are ever the same. Had I not changed careers, I would never have realised this opportunity.

To be in tech, you do not need to be a fully fledged coder, or even a technical person who loves science, engineering or maths. A degree is not the only route; other options include apprenticeships, work experience, short-term contracts and government-funded schemes. Tech incorporates many different parts of the business world and many different roles; I will cover some common and typical technology roles in detail later in the book.

Tech is such a wide area, I wonder if school curricula or careers advisors are doing enough to encourage girls to consider it as an option consistently across education. As this is such a crucial industry which currently lacks female input, Dr Anne-Marie Imafidon MBE co-founded Stemettes, an award-winning social enterprise working across the UK and Ireland. It supports and inspires girls, young women and non-binary people into STEM subjects. Over 50,000 young talent have attended events, workshops and experiences for free.

I have included a link in the Resources section at the end of the book if you want to know more.

What you need for a career in tech

Without even realising it, you may possess talents, skills, experience and energy that the tech industry desperately needs. We dedicate at least a third of our working lives to our career, so we should ensure that it is a pleasant and rewarding experience, while those who want it can make a good living and climb the ladder.

However, being an in-demand and respected tech professional takes more than just technical skills. To be clear, having good technical skills is always a benefit, and depending on where you land and which role you are interested in, to know a little coding or use of various programming languages may be required. Knowledge of coding or programming languages can help you in careers that analyse large amounts of data or keep computers and systems or networks secure. Some roles require an understanding of complex technical concepts in software development.

Aside from that, though, tech professionals need to be able to communicate effectively in both oral and written forms. They must be problem solvers, treat challenges as opportunities and know how to navigate obstacles and move forward when things get tricky. They need to have a continuous growth mindset, understanding that learning is a lifelong process, rather than having an 'I went to college/university' or 'I completed an apprenticeship, so that is enough' attitude.

To be successful, you must be focused, talented and knowledgeable. Good tech professionals know that a massive contributor to their success is a plan. A promising career cannot materialise without a plan. When you know that you need to make a career change, perhaps look at my blueprint – my story – as a guide on your journey. You can, of course, modify it as necessary to suit your circumstances.

In later chapters, I will help you to think about what is available in the tech industry and how you can pivot/change. Whether you are thinking about your first career or figuring out your next meaningful move, you will need to understand what a new role may look like. Included are tips and shared experiences to help you get the skills to succeed in the position you choose and stand out in the crowd. I will share my journey in corporate and include interviews with tech professionals who give colour by sharing their stories, successes and lessons learned, all with the challenges of women in mind.

Women's challenges have been repeatedly written about, documented, analysed and studied for years, and rightly so. Now more than ever, we need to see women of all ages getting into and succeeding within tech, and growing our pipeline of role models.

The Tech Talent Charter, founded by several organisations across the recruitment, tech and social enterprise fields, has been supported by the UK Digital Strategy since 2017 ('Tech Talent Charter – Diversity in Tech' 2023). It is committed to securing the future of the tech talent pipeline. A key driver of its annual 'Diversity in Tech' report is showcasing DE&I data from its signatory base of companies with tech needs in the United Kingdom. This report exposes some interesting data that accentuates the facts, and is worth a read. If you are interested to know more, you will find the full reference at the end of the book.

The ONS and the UK Census 2020 stated that almost all people aged sixteen to forty-four years in the UK at that time were recent internet users (99%), compared with 54% of adults aged seventy-five years and over. It is interesting to note that there was little change in internet use for people aged sixteen to forty-four years between 2013 and 2020, but the proportion of users aged seventy-five years and over nearly doubled, from 29%. This is just one example of a wealth of evidence that technology is not only growing, it's here to stay. The sooner women are aware of the opportunities on offer, the better it will be for us all.

In 2021, the WeAreTheCity Tech Women network partnered with IPSOS Mori and the Tech Talent Charter to examine barriers to women in the industry ('Tech Talent Charter – Diversity in Tech Report' 2021). At the time, only 21% of people working in the tech industry were women; an astonishing 84% of survey respondents saw salary as the primary driver of women joining a tech organisation.

Undoubtedly, salaries can be lucrative in this well-paid industry with a vast range of dynamic, exciting and ever-changing jobs, as tech touches every area of our lives. Sadly, many of us suffer from a lack of confidence or have difficulty stepping out of our comfort zone, and I am no exception. This is why I – mistakenly – did not change careers sooner.

My interest in science could have been better. I did not enjoy it at school and thrived more in the arts subjects, but being proficient in science and maths is not the be-all and end-all requirement in tech. Simply put, we need diverse representation at all levels to address the current and future technological challenges.

Dispelling myths – To code or not to code

We saw proof that tech facilitates many livelihoods during the Covid pandemic. The People and Skills Report 2022 from Tech Nation says, 'During the uncertain times faced by most people over… the Covid pandemic, technology has been an enabler for individuals, companies and communities. It has facilitated new ways of working and kept the economy buoyant.'

Thanks to technology, we now experience better ways of operating professionally with the flexibility of remote working. This has helped many of us meet other priorities in our daily lives, such as being parents, caring for a loved one or simply fitting wellbeing in with better fitness routines. There are plenty of people who could never have imagined being able to work like this pre-pandemic, and as we all know, it was only possible with good tech and support.

The People and Skills Report 2022 goes on to share, 'More than two million tech vacancies were advertised over the last year, more than any other area of the UK labour market'. According to the report, tech vacancies overtook trade and construction, teaching and healthcare roles.

With such a shortage of skills in the tech industry, what is stopping people, especially women and girls, entering this exciting sector? There is a widely held belief that one must study and excel at programming or become a coder or programmer to have a successful career in tech. This is not true, so do not be put off.

Understanding programming or knowing how to code in languages like Java are good skills to have as they enhance your understanding of logical computer-like thinking, for example, how computers break down problems and solve them. This can aid in developing your problem-solving abilities and knowledge of how applications work. It can also enhance your understanding of how coders translate and approach problems.

If you are interested, I encourage you to get a taste of programming or coding via a mini course, boot camp, hackathon or online class with self-study. Many communities and organisations hold day-long introduction sessions for coding. These days, there are so many ways to find yourself a free taster – just for your own satisfaction.

Please note, though, that this *does not* mean programming or coding is your only way into working in tech. Coding or programming can be like Marmite – you either love it or hate it! I will cover more about the difference of being a coder and programmer in the Tech Trades chapter. It is not for everyone. However, it is beneficial to go through some learning to know what it involves and to get a flavour of it.

Just like a recipe when you are cooking, the tech sector has many ingredients. It is so variable with its roles and areas, so be aware, as when you are making a pie or a cake, that some ingredients should be a part of your journey and some just won't be to your taste. It's all about personal preference.

I did not enjoy creating code and learning programming during my MSc degree, but I quickly realised through studying Java and Structured Query Language (SQL), which is used for databases, that I am drawn to the business or analytical side of the tech sector. This involves softer and less hands-on technical skills, for instance, communication, presenting, meetings, problem solving, analysis and dealing with people in general.

That said, I did learn SQL, Java and hypertext markup language (HTML) skills used for building websites. Some, like HTML, are quite simple to learn, while Java reminded me of knitting in that it was more intense, and if you get one stitch wrong, it messes up the whole garment. For example, a scarf ends up with a small hole showing, and then you have to go back to correct a whole row of stitches. I tried knitting for a short time as a girl, when my late mother created a beautiful crocheted garment for me, which encouraged me to want to try basic knitting, but my enthusiasm did not last for long. It was the same in coding. It tested my patience as I had to go back if the program was not working and read every single line of code meticulously, each letter in the script to find out where the mistake was before I could move on.

My programming course was a good one covering practical areas, and the first primary assignment I completed was to create a telephone directory. The main observation I had, as I had never coded or written computer programs before, was that I couldn't believe the number of lines of coding required to ask the computer to perform a simple task, like finding the telephone area code for London, Leeds or Bedford. All to receive a three- to five-digit number! It was a great learning experience for me and I passed my end-of-year exam with a good grade, but it was a journey which I didn't enjoy.

However, this was balanced out with other subjects I did enjoy, such as business information systems and systems analysis, which brought out the artistic side in me. I also loved the creativity in internet and web design.

Why have I shared my example of learning coding if you don't need it to enter the tech industry? It's because I recommend that you at least try it. This general information could be something you need to know about and help you understand how things work in the background. If, for example, you go to work in a company with end-to-end development of applications, like banking software or fintech, it is helpful to know a little about how code works, and once you learn one program, you'll find they all have similar principles.

If you enjoy programming or coding, that is OK – do keep it up! If you do not, this is also OK. You will find that soft and non-technical skills can secure you a good job within tech. I am not an expert in every area; I'm simply sharing insights so you can be more aware of what is available to you, and then you can seek expert help in any decisions you make or to allow you to learn.

Where are tech careers?

Which companies do I start with? These days, *all* companies use technology and have tech-related jobs, not only the world's biggest tech firms like Meta, Google or Amazon. You need to think broadly about an industry you may be interested in. Many companies require tech support, although if you're just starting out, perhaps find a position that provides the experience you want and teaches the skills you'll require to advance.

Financial services, such as online banking, that need 24/7 customer access make significant investments in their tech and infrastructure. Even if large companies off-shore their call centres, there is still a need locally to meet demands. Mobile banking is big, so just think of the number of people required to develop these applications. The team can include coders to work on the developments and testers for the software, customer-facing staff, analysts to ensure requirements are met and people who work on security to ensure the safety of the information is high. As you can see, there are several roles involved in a tech project, and this example is just in one industry.

Health is another good example. Healthcare experts use and develop all kinds of technology to improve treatments; you may have noticed that electronic prescriptions have been in use for some time. The National Health Service (NHS) is one of the largest employers in the UK, and many branches have tech departments. This is where I started when I changed career, and it could be a good place for you to start too. Not only will you be working for a highly respected and essential organisation, but you will find the NHS has a great working culture.

You can search for companies you may be interested in working for on the internet and look at their job boards. Better still, networking will help you learn more about the industry trends. We will cover networking in detail in a later chapter.

Who can go into tech?

The simple answer is, anyone can. You do not need a degree in tech. We all have different circumstances; it is excellent if you want to study, but it is not required as a must. A degree can provide a great foundation, especially if, like me, you are completely new to tech. However, the valuable learning happens when you are working on the job.

Some of the most famous innovators came from little or no education. Think of Mark Zuckerberg, the founder of Facebook (now Meta), or Thomas Edison, the inventor of the lightbulb, both of whom dropped out of school. Microsoft founder Bill Gates spent two years at Harvard, but still needs to complete his degree.

You can see from the statistics I have shared in this chapter that opportunities await. Far from having to be a coding expert or have a degree in computer science, you will find the potential to join the tech industry is open to all. There are many paths into the sector and a variety of skills are required.

This chapter has set the scene for you so you know why there's a need for more diversity in the tech industry. The talent shortage in tech means there will always be a demand for employees. We have covered the basics of bias in tech developments and I have shared the start of my own journey.

In the next chapter, I will share what I did to make my move into the technology industry, the research this involved and what the course entailed. You will also learn from a successful woman about what inspired her to work in technology and how she came to be in her role, and gain insight from a data analyst who did not come from a tech background.

TWO

Tech Trek: Embarking On A Journey Of Boundless Opportunities

> 'I've always been more interested in the future than in the past.'
>
> —Grace Hopper

Grace Hopper was a famous computer scientist and a pioneer in the early days of programming. She helped in the development of Common Business Oriented Language (COBOL), and although this has now been replaced with alternatives like Python, Java, JavaScript or Cobalt, it was essential for business computer programs in finance and HR. Born in 1906, Grace played a crucial role in early technological developments. Her work spanned many decades and her influence continues to be celebrated and inspire women in tech.

Just like Grace, we all need to be interested in the future, and what better place to start than in the tech industry? Tech epitomises everything that is exciting about the future.

In this chapter, I discuss the many ways to enter or pivot into a tech career and be successful. As we have seen in Chapter 1, there is no one path because each individual will have a different journey and circumstances. What I share in this book is either what has been most helpful to me or based on my observations of others and the experiences they share. My aim is to give you insights into what I've learned through my networks and events, conferences and from speaking to many people over the years.

My journey to a different path

When I changed career, leaving a perfectly good job in a senior position, people inevitably made comments and judgements about my choice. It does take confidence to make the move, but only I knew what was right for me, and the same is true for you. We are all entitled to make a change for the better, even if other people disagree with our decisions or choices.

When I changed my focus mid-career, there were people who thought they knew better than I did. However, I persisted with my goal and I never looked back. 'Who cares?' I asked when they tried to sway me with their opinions. Who could have guessed that my income would more than double?

Having decided to change my career, I took a year out to study computing and IT. Although I had worked my way up in my previous career, I knew I needed a more fulfilling job to maintain my interest and inspire my learning – a job that also had the potential to increase my salary. My old job as an assistant manager in a large corporate had become mundane. I had moved around many office roles and felt I had come to the end of the road.

Although I was a manager running a team with serious responsibilities and was competent at my job, and my annual performance reviews were always good, I felt I had a limited scope to learn, progress and earn more for my worth. However, my main driver for change was that the job did not keep my attention and interest. Mistakenly thinking it was the best option, I made my first career change within the same big corporate, into a different area.

I soon realised I was doing the same thing and expecting different results, and then I questioned why. After that, I was determined to change companies, thinking this was the answer and all would be better. Again, though, I was unsatisfied, and again I questioned why I had gone from one unfulfilling role to something similar. It became obvious that to be happy and more fulfilled, I needed a complete pivot, although at the time, I did not know what contributed to my unhappiness at work.

I was drawn to the tech industry simply because it is dynamic. It has extraordinary variety and is full of opportunity. I enjoy reading and keeping up to date with news and changing technologies, but as someone who did alright but never really excelled in science and maths-related subjects at school, I did wonder how on earth my interest alone would help me land a successful career in technology. At this time, I had no formal technology training, so I was unaware of the options.

I spent a few months looking at options, getting answers to my questions by evaluating my strengths, weaknesses, experience and current skills. I researched in depth the potential of changing career into tech, which involved me speaking to people in the industry, networking where possible and learning as much as I could about the latest developments and what was up and coming.

Interestingly, I approached the Chamber of Commerce and conversed with a career guidance official there. During my school and college days, I'd never thought much of the standard of advice from career guidance experts, and unfortunately, my experience as an adult merely confirmed that belief. In fairness to the advisor, the tech industry is so vast and quickly changing, their knowledge of it would constantly need to be updated, but they could only comment on coding. New roles are always coming out and job names are changing, so it takes work to keep up knowing about the skills required in specific IT roles.

Nevertheless, I landed on the decision to take a risk and study – despite my prior knowledge of tech being limited to the use of Microsoft Word, Excel and PowerPoint. I wanted to give myself the confidence to transition and gain practical and theoretical experience from the course, so I chose carefully.

My career transformation in tech

I had completed my first degree, a BA Hons in business and finance, by studying part time in the evenings while working full time. This time, I took the plunge to leave my job and complete a postgraduate course full time, so I could start working on making the change quickly after finishing.

I researched in depth what was available as there are many technical courses out there. The option I chose was a one-year intensive MSc conversion course in computing and IT, a popular course as it had as much practical work as theory. This covered all the areas across the spectrum with the practical learning within the course modules, and was enough to leave me with an excellent all-around knowledge of the industry.

It also gave me the comfort of knowing what I enjoyed and what I would be good at and not so good at.

If you are changing career, regardless of the approach you choose to transition, ensure there is enough practical work in your training, whether that's via a course or on the job, for you to get a feel of what it is all about. Pivoting in your career is a risk, so you need to weigh up, as I did, what the worst is that could happen.

My answer was: 'To fail or not secure a job in the tech industry.' I had my work experience to fall back on, so the worst-case scenario would be to go back to work in the same industry as I had previously left. At best, I would secure a new job which would entail working my way up and learning within an industry I longed to be in.

I love the arts and writing, but I knew a writing career could be precarious, and most writers earn little. I also love public speaking, but that industry was not as big then as it is now. At least, it wasn't big enough for me to pursue a new career in it with confidence.

The computing course gave me a broad understanding of the tech industry and was ideal for someone so blue and new. It included modules on Java programming, networking, basic computer fundamentals (hardware, etc), systems analysis, business information systems and SQL language with database knowledge, which really helped me learn. I quickly became conscious that coding was not an area I could do full time as

I like dealing with people and using my soft skills. Also, I talk far too much to sit quietly in front of a screen most of the day! I would never be able to get pages of code done in time.

All went well during that year of study, even though it was intense, especially as I had to learn everything from scratch. I gained excellent grades and was excited to start work in this new world. During the summer period of that year, before I graduated, I took the initiative to ask the faculty head at the university if they had any work in their tech department. To my surprise, they came back to ask which area I was interested in, and I gave a couple of options based on what I had enjoyed during the course: information systems analysis, information management systems and networking.

I spent around two-and-a-half months working in the networking department of the university, which gave me experience of the basics of this area. The system that the networking department team used at the time to run their computers was called Novell, which is old and rarely used now, but the experience gave me knowledge of practical concepts. I was also pleased to have something relevant to the tech industry to put on my curriculum vitae, and not only that, earning a salary after having been out of full-time work for a while was a gift, too.

The transition from working to not working had been a challenging one. For me, it took a lot of planning my finances and cutting down on visits to shops for new outfits (sorry, Zara and Mango). Plus, I missed many socials with friends during that year of intensive studying.

As I was close to finishing the course, in the run-up to handing in my dissertation, I remember sitting in my room, busily doing the research pieces for my write-up of my project. The weight was about to be lifted from my shoulders. This was the final push, and I was looking forward to graduation and meeting up with fellow students again afterwards.

I came running downstairs from my room to take a break and watch the news on television. The day was dreary, dull and rainy, and as I sat down, I heard the word 'technology' being mentioned by the newscaster.

I felt a sudden surge of panic. The economy had been booming before I joined the MSc course. The tech industry had been creating many jobs; people were making so much money in the excitement of new companies being formed. Now, in contrast, we were witnessing a great big dip or crash in the economy. Suddenly, it was boom to bust.

I froze. 'This cannot be happening,' I said to myself. 'I need to find a full-time job as soon as I graduate.'

When considering my career change, I had thought of the risk that I was taking. Plan B was always in my back pocket – to return to a job similar to the one I had left, but this was not what I wanted to do, so it was not at the forefront of my mind.

Things do not always work out the way you plan them. After applying unsuccessfully for many tech roles, which was quite demoralising, I was still glad to have completed the course. To keep myself busy, I found an opportunity to do some voluntary work at my local hospital in the tech department. Coincidentally, I had interviewed the head of tech there as part of my dissertation, so was not going in entirely cold. It was a good idea for me as I knew I had to start somewhere.

This experience gave me a good grounding as I worked on the tech department's helpdesk and shadowed a technician who had excellent skills and explained so much to me. I found myself fixing PCs or faulty scanners, along with rolling out a new scanner system across the hospital as at the time, they were fairly new and were to be used to scan patient files for data.

Often, I would find myself on the floor, reaching for wires and cables, a far cry from my formal office days. I'd wear trousers at work or even adopt the casual nerdy tech look, which was so different to what I had been used to, it was fun. Putting myself into tasks I would never have dreamed of doing before, but which would become routine, was quite an eye-opener for me. It taught me to be more adaptable to ever-changing requirements and to make the best of being a trendy female tech geek.

Eventually, as I had been interested in all the hospital software and how the IT Training Head and the team conducted their training, I was allowed to do some training myself and carry out some sessions, which were well received. As a result of all the experience I gained during this time, I secured my first job in a London NHS Trust in the IT training department.

Here is the thing: you have to start somewhere. Although I was no longer in charge of a team and had to learn from the ground up, I was prepared for this as I knew I was making a complete change, and was utterly open to learning and trying everything. I eventually got to where I wanted to be, but my journey took a few unexpected twists and turns.

The NHS Trust in London hired me to be an IT trainer for its patient-systems to train staff on how to use the systems. In this particular NHS Trust this department was new and training rooms had to be set up from scratch, so this was how my journey into project management began. As I thoroughly enjoyed this, I went into coordination in tech projects, and then joined the National IT programme, and so the journey continued.

I have never looked back. From the NHS, I went into telecoms, IT services companies, and oil and gas, through to the present where I have spent a lot of years in financial services.

What has inspired other women to work in the tech industry?

Let's now hear from Kasia Wojciechowska, Regional Head of Client Accounts and Lead Neurodiversity Employee Resource Group at Peregrine (formerly Capita).

INTERVIEW WITH A PROFESSIONAL – KASIA WOJCIECHOWSKA

'I achieved a master's degree in communication and management of information resources. While I was not committed to pursuing a hands-on IT career, my passion for technology developed around human interaction with fast-developing technological advancements. During the internet era of the early 2000s, the development of e-commerce, social media and portable devices, I knew that working in tech was where I wanted to be.

'I was fortunate to meet a chief executive officer of a fast-growing professional and tech services business in my mid-twenties, who opened the door to my first corporate career. All of a sudden, I had to learn a brand-new world of IT, the seven phases of the software development life cycle (SDLC) and the basics of how financial services and banking worked. I was thrown into the deep end, supplying tech skills to the top investment banking clients, including UBS, Credit Suisse, Citi and Barclays, to name a few. Being able to work with both technology leaders and senior decision makers in this space globally has given me a fantastic insight into corporate tech.'

Another successful woman working as a cyber security consultant told me:

> 'My inspiration for joining tech came from being an introverted individual with a deep passion for data. I love data analysis, problem solving and the continuous learning opportunities of tech, allowing me to explore diverse roles. Initially, I wasn't fully aware of the potential remuneration, but I am pleased with the rewarding career path tech has offered me. I encourage others to consider the tech industry for its dynamic nature, endless possibilities to learn and grow, and the chance to make a meaningful impact.'

A highly skilled data analyst with a background in HR said of her journey that she developed a passion for working with data. She told me:

> 'Data is like a puzzle that you can piece together to bridge the gap to help any organisation thrive. Going into tech may not always be easy, but it is down to you as an individual to make your journey easier. I went into tech with no previous background, but I used my transferable skills from my previous HR roles.'

This woman has been recognised for her excellent research skills, using limited amounts of data to spot important trends and patterns in companies' recruitment processes.

My advice to you is:

- Be prepared to take risks, which can pay off, but weigh the pros and cons before embarking on your plan.
- Adapt along the way, as no career trajectory is a straight line. There is no standard route into a tech career.
- Keep up with learning new skills as change happens a lot in this industry. Those who put in the time and effort to continuously improve do well in their careers.
- Have patience. Initially, you may struggle when learning new skills as it takes time, so embrace the change and take the long-term view. Avoid shortcuts and make sure you understand your new learnings correctly.

After speaking to this lady, my personal experience is testimony to the fact that there is no straight route into tech. We set out for one thing and adapt along the way. As the excellent book *The Squiggly Career* by Helen Tupper tells us, it's time to ditch the concept of the career ladder (Tupper, Ellis, 2020). You may need to make deviations or take different routes until you reach your preferred destination. Learning is a continual journey.

In this chapter, I have shared my own story of getting into tech. This is what worked for me; it does not mean you need to take a long course of study. There are many different ways to enter tech, so choose one that works for you. You have seen other opportunities available through the stories my guest interviewees have shared with us about how they entered tech, often with little experience. Anything is possible if you put your mind to it.

In the next chapter, we'll investigate some of the current technologies to bring you up to date, but please note these are by no means all of them. That would fill the whole book! Please also note that these technologies will advance rapidly, so if one takes your interest, I advise you to make sure you stay up to date with it. I have selected some that may spark your curiosity, such as artificial intelligence, crypto and blockchain.

THREE

Tech Trends: Unleashing The Future – A Glimpse Into Emerging Technologies

> 'Nothing in life is to be feared, it is only to be understood. Now is the time to understand more so that we may fear less.'
>
> —Marie Curie

This chapter will familiarise you with some of the keywords and areas of emerging tech, as well as the popular trends already with us. There are so many emerging technologies that I could not possibly cover all of them, so I have picked the most topical at the time of writing.

There is nothing to fear from tech, but to mitigate any risk of fear, tech does need to be understood by all. As the great Marie Curie suggests in the quote that opens the chapter, if we understand it, we won't fear using it. Perhaps more importantly, with understanding, we can ask the right questions.

I was once asked to chair a panel for a session at a women in tech conference, and I remember the subject being too technical for the mainstream audience as the particular technology was relatively new. When I attended the pre-meeting, it was evident that the people on the panel were all close and familiar with every piece of technical jargon on the topic, but they had clearly forgotten the audience would be a mixed range of levels when they had prepared in advance all the questions I needed to ask them, along with their answers. Some attendees would be familiar with the subject, some would not, and many would be totally new to it. Some might have heard of the terms, but not understood them at all.

Why am I sharing this? It is crucial to be aware of who you're speaking to during conversations about tech and amend your language to the right level. Too often, the experts forget that not everyone will understand certain concepts or acronyms, or even what the subject is all about, so it might sound like a foreign language to the listener. This merely exacerbates any fear the listener may have around the world of tech, which is not what we want at all.

This is why I have devoted a whole chapter to the subject of emerging technologies, explaining the examples I have chosen in a simple way. Hopefully, this will pique your interest. Then it is over to you to research your particular areas of interest, keep up to date with the rapid advancements to come and, of course, share your learnings with others. First, though, I recommend that you familiarise yourself with the basics, which will spark your curiosity to research further.

Before we start, let's have a look at what exactly an emerging technology is. Quite simply, it is a technology that exists, but we have yet to experience its true potential. Further refinement or research is required before we see it in common use into the future.

Things like self-service checkouts or contactless payments with our credit or debit cards, which we now take for granted, were once emerging technologies. At that time, these mainstream facilities were not mainstream. If we go further back in time, the same applies to email. When it first came out, many people didn't use it initially as they were still used to 'snail mail'. It took a while for it to become the norm, and now we can't imagine life without it.

Now let's take a detailed look at three fascinating examples of technologies that are emerging at the time of writing.

Your path to understanding artificial intelligence (AI)

You have almost certainly seen all the hype in the media and news about AI, as this is exploding across the tech world. None of us can go anywhere without hearing about it. This is why I have chosen AI as the first emerging technology to cover in this chapter.

AI is a term for a computer system trained to do tasks that traditionally humans would have done in the past. It imitates and learns from the level of intelligence of humans so it can make decisions like a human. In essence, AI is the simulation of human intelligence processes by computer systems.

Until recently (at the time of writing), AI was not a mainstream topic, yet it has been with us in different forms for years. Now that it is developing fast, the conversation has started with a vengeance. The tech giants are competing hard to bring out the ultimate AI products that will end up being the ones we all want to use.

It is said that AI tools will make our lives easier. I agree with this as AI means we will have more time to do the things we like or the important work that software has limits in doing, such as supporting the elderly, caring for the sick and teaching the young. There is one caveat, though; emerging technology must be used appropriately. We will discuss the ethical issues and regulation of AI later in this chapter.

AI has set in motion a huge debate. As ex-Google officer Mo Gawdat said in his interview with podcaster Steven Bartlett, 'It's the most existential debate and challenge humanity will ever face. This is bigger than climate change, way bigger than Covid.' (Bartlett, 2023)

AI will be a truly transformative technology for us, if used and developed in the right way. It will reshape the way things are done, so everyone who is looking into this, including big tech corporates, governments and regulators, needs to be reacting fast. They must also be proactive to ensure everyone can be in a position to understand this tech and use it ethically and wisely.

AI is already in our lives. I am sure you have heard of Amazon's Alexa; this is a simple form of AI. Computers can be taught how to analyse information and draw extrapolation from patterns within datasets. The more data we give them as they cope with vast amounts of information, the better they should get at this.

Let me elaborate further. There are different terms used by people when they refer to AI. At the basic level, we have simple AI. We also have more advanced technologies, such as machine learning (ML) and deep learning. Each has a specific meaning, so let's cover them one at a time.

What is simple AI?

A chatbot is the simplest example of AI. Think about when you go online or make a phone call requesting help with a query, and a chatbot responds either through voice or text or online chat. It is used to decipher where you need to go with your query.

For example, you may have interacted with chatbots on the websites of telecoms or energy companies, or over the phone on customer service or shopping lines. These chatbots must be programmed by humans to work, so the programmer needs to know all the different types of information the chatbot requires, such as tasks to complete. The programmer then develops scripts and prompts that the chatbot uses, all based on this information.

As the end user, you will recognise that you are responding to pre-programmed scripts, because if you ask anything that's unclear or something that the chatbot has not been programmed to recognise, it will not respond how you want it to. In other words, it won't answer your question, which can be quite frustrating, as it is only pre-programmed to answer certain queries.

To summarise, in simple AI, everything is programmed by humans, and the AI cannot make adjustments or decisions on its own. Only the coder or programmer can step in to add any new requests. This is why chatbots do not have the ability to learn and adapt.

What is machine learning (ML)?

ML is where the technology learns over time from data or inputs, and then refines its decision-making process in a creative way. For example, when a chatbot equipped with ML receives a request or question it does not understand, it learns from the question it is asked and accesses all the data and algorithms at its disposal. An algorithm, put simply, is a series of instructions that are followed to produce an output. You, the end user, then tell the chatbot if the answer (the output) was helpful or if it has not correctly answered your question. Over time, the ML chatbot learns, and the next time it is asked, it will offer a better and hopefully correct answer.

No further programming is required with ML as it does everything independently. Think of Amazon Alexa or Siri on an Apple iPhone as an example. The most successful versions of ML in recent years have used a system known as a neural network, which is modelled on how scientists believe the brain works at a simple level.

What is deep learning?

Deep learning technology takes things much further. It has sophisticated algorithms or neural networks that imitate or mimic a human brain and are used to complete a task. This means deep learning requires large amounts of data and power.

A good example of deep learning is self-driving or autonomous cars. There may even be hope for those of us who don't love housework and tedious chores that we will soon have technology to do them for us.

ChatGPT

You may have heard of ChatGPT – the tool that brought AI technology into mainstream awareness. *ChatGPT* is an application that you can download online and ask almost anything. It's just like Google, but better, as a chatbot quickly responds in a conversational style to any user asking a question.

These chatbots are trained to use a vast amount of information from the internet. ChatGPT can even take the grind or toil out of tedious jobs to allow people to do the things they enjoy and like most. For example, it aids students with homework (probably much to the dismay of teachers), can write blogs and helps people in business access information based on data that it has been given.

Emily Bell from the *Guardian* newspaper concludes that we need to be wary of using this AI application for articles, homework or blogs, as checking and human editing may be required (Bell, 2023). In her article, she referred to Felix Simon, a communications scholar who interviewed 150 journalists and news publishers in relation to AI in newsrooms. He said that ChatGPT could aid journalists in transcribing interviews or absorbing datasets, but accuracy, bias and data provenance would still have to be checked by a human being. This illustrates really well what I am going to share with you.

Think about what you are using ChatGPT for, as it has been programmed from vast amounts of data that are not necessarily correct or relevant to your request. Always make sure you include your own analysis, particularly if you are using it for homework, blogs or business. Especially for homework, as teachers will reward you for original thinking and recommendations, while AI has the potential to spew out inaccuracies and make your content meaningless. Question everything you read online – in the same way you may have questioned your parents when you were a child being told wrong from right.

There can be much misinformation online. You may have heard the expression *garbage in, garbage out.* If ChatGPT uses incorrect information, the output will be garbage.

AI can, however, save us all time and effort, and there are examples where it works really well in creating images, creating music, computer programming and even personal health treatment plans. I would love AI robots to complete all my housework; that would be an excellent time saver!

Although people are saying AI will take jobs away from humans, it is no different to other periods of innovation in history. For example, when automation took jobs from factory workers in the 1970s, there was a big shift in labour practices with people moving into service industries. Progress always disrupts something. Even though some jobs may go, AI will create new ones. AI can even enhance jobs.

The World Economic Forum 'Future of Jobs Insight Report' 2023 predicts that a number of technologies including AI '…are all expected to result in significant labour-market disruption, with substantial proportions of companies forecasting job displacement in their organizations, offset by job growth elsewhere to result in a net positive.' That may well happen, but think about all the new jobs, mainly related to technology, that will appear on the market. It is good to see that the loss of old jobs will be offset by new ones, which indicates positive signs. More and more jobs will require humans to work effectively with technology.

INTERVIEW WITH A PROFESSIONAL – KASIA WOJCIECHOWSKA

I asked Kasia Wojciechowska to tell me what she thinks about AI. This is what she replied:

'AI disciplines, such as natural language processing (NLP) and ML, play a central role in many tech or large enterprise businesses now. Generative AI tools are already being used to boost productivity and offer more innovative ways of working. The conversational AI's global market is expected to grow and interactive voice response (IVR) is growing too, all because they can work with copious amounts of data. It is truly positive to see UNESCO issuing its first global guidance on GenAI and its ecosystem in education and research.

'There is certainly a promising future ahead. While speed and scale are at the frontier of businesses, AI cannot create inclusive cultures, collaboration or launch new ideas. It doesn't invent things by itself or develop new products or services that meet people's changing needs. Therefore, let's embrace it, but with caution.'

If you are curious about AI, keep learning, researching and stay on top of the trends. This will help you to become aware of the next big thing, and you are never likely to fall behind or be out of a job.

In general, tech changes so fast, none of us can keep up with everything. As you are reading this, AI will be continuing to advance.

Cryptocurrencies

Cryptocurrency or crypto is a type of digital money that uses special techniques to keep it secure. Imagine you have a message you want to send your best friend without anyone else understanding it. That's where encryption comes in. Encryption is like a secret code only you and your friend can understand. Crypto uses encryption to protect its messages and ensure they can only be read by the intended recipient.

Encryption takes a crypto message and scrambles it up in a unique way. When the intended recipient receives the scrambled message, they use a unique 'key' to unscramble and read it. Here's the really cool part with crypto: you don't need someone else to keep the key and unscramble the message for you. Instead, the key is a unique password which allows you to send each other messages securely without anyone else being able to read them.

In simple terms, crypto is like having a secret language. It's a way to send messages or trade digital money securely without anyone else snooping around. When we talk about trading digital money securely, we are referring to buying, selling or exchanging cryptocurrencies safely and securely.

Cryptocurrencies, for example Bitcoin or Ethereum, are online forms of digital money. Bitcoin, created in 2009, was the first cryptocurrency and is an officially regulated currency in El Salvador, but there have been thousands of cryptocurrencies created since then, each with its unique purposes.

Just like you may have traded toys or stickers with your friends when you were a child, you can trade cryptocurrencies with other people, but unlike physical goods, cryptocurrencies are purely digital. Their ownership is recorded on a particular type of technology called a blockchain (more on blockchain technology in the next section). The 'secure' part comes into play because cryptocurrencies use encryption and the blockchain to ensure that these trades are trustworthy and can't be easily tampered with.

When you want to trade cryptocurrencies with someone, you both need to have unique digital wallets. These wallets are like your personal online piggy banks for your cryptocurrencies. Each wallet has a unique address, like your home address, to send and receive digital money – cryptocurrency. The transaction is recorded on the blockchain, which in this respect is like a public ledger or a giant digital record book that keeps track of all the transactions with each specific cryptocurrency.

The blockchain is secure because it's not controlled by just one person or organisation. It's distributed across many computers worldwide, making cheating or manipulating transactions difficult. This secure system allows people to confidently trade cryptocurrencies, knowing that the transactions are being recorded accurately and can't easily be changed or hacked.

To summarise, when we say that we trade digital money securely, we are using encryption and blockchain technology to ensure that when we buy, sell or exchange cryptocurrencies, it happens safely and in a trustworthy way. This is because it is difficult for unauthorised parties to gain access to encrypted information as anyone without the secret key is unable to interpret the information.

Decentralisation

The potential impact of crypto is that once people are more educated and comfortable with the technology, it could change how they engage with financial systems. Unlike traditional currencies, cryptocurrencies are not controlled by any central authority, such as governments or banks. For example, traditional banking and financial services in the UK have two regulators primarily responsible for their authorisation and supervision: the Prudential Regulation Authority (PRA) (regulates and supervises financial services firms) and the Financial Conduct Authority (FCA) (which is an independent non-governmental body).

Crypto is helpful for people who lack access to traditional financial services. For example, UNICEF uses Bitcoin to get money into environments where children need support and other means are unavailable.

It is also an excellent avenue for inclusivity regarding finance and empowerment. Think of the problems in some countries where there is such poverty that people cannot trade their currency. They will no longer need to rely on intermediaries to transfer or manage money. As this technology is global, it is accessible to all and used by anyone; all you need is an internet connection.

Advantages and risks of crypto

Advantages include:

- **Financial inclusion**: crypto gives greater empowerment to those who cannot easily engage with financial systems as they have direct control over their funds. They are not relying on banks or other intermediaries for the management and transfer of monies.
- **Costs**: with no intermediaries, there is a reduction of costs.
- **Global reach**: as crypto is global, financial transfers are less time-consuming than traditional methods, which is good for businesses.
- **Security**: as there are encryption techniques, transactions are secure.
- **Innovation**: crypto will potentially open up the innovation of financial products, such as smart contracts, decentralised applications and more.

As with all new technology, there are risks and challenges related to crypto. These include:

- **Volatility**: every investor in stocks and shares knows that there can be a lot of price volatility, but with crypto, this risk is even higher.
- **Uncertainty**: there are regulatory uncertainties as new frameworks are still being established.
- **Lack of understanding**: crypto is complicated, so I strongly recommend you only invest once you have researched it thoroughly and spoken to those who already use it.

Blockchain

Dr Anne-Marie Imafidon MBE describes blockchain in her book *She's In CTRL* as 'a set of records of any kind of transactions (they do not have to be financial) that is kept publicly so that they cannot be tampered with or edited' (Imafidon, 2022). In other words, it stores transactions and can be used to track nearly anything.

Blockchain is a database that is secure. Like a ledger shared by two or more parties, it holds the information on a transaction and is used to exchange an asset (which could be money or deeds or supplies, even leases to property) from one party to the other. Once the transaction information is entered into the blockchain, it is time-stamped into the ledger and cannot be altered by anyone.

A blockchain is suitable for use by businesses as it allows them to monitor from start to finish where any goods are and if the goods are genuine. Financial services can send payments faster than by traditional methods and reduce the fees they charge to customers; in addition, fraud and overall risks can be reduced.

All the records held in a blockchain remind me of a complex chessboard because of the way it all works. It is comprised of many blocks of information, all chained together and unbreakable. The digital currency is stored on the blockchain, and its value fluctuates based on supply and demand.

When you place an order with Amazon, you know how the item reaches you, and that the people who supplied Amazon will get paid, because you know it is being purchased from a reputable source. Blockchain is a series of similar events, and these blocks of information held in the chain are not editable or able to be tampered with in any way. This is why it is used for money – the charity UNICEF uses blockchain to track how its donations are spent.

Advantages and disadvantages of blockchain

The major advantage of blockchain is that it is easy to verify transactions as it is tamper-proof. It offers greater security than other platforms, even though it does not guarantee that it is entirely safe. Clear audit records are held, and it has reduced the time and the costs involved in making a transaction or transfer.

The main disadvantage is that it is relatively new and has yet to enjoy mainstream adoption, so there is a lack of standardisation in its use, not to mention its understanding. Also, blockchain networks can be slow due to the high computational requirements needed to validate transactions, because they have so many of them. This is unlike traditional payment systems.

As the number of users, transactions and applications increases, the ability of blockchain networks to process and validate transactions in a timely way becomes more and more strained. This makes blockchain networks challenging to use in applications requiring fast transaction processing speeds, so scalability must be looked at over time.

The process of validating transactions on a blockchain network requires a lot of computing power, which in turn requires a lot of energy. This has led to concerns about carbon emissions and the environmental impact of blockchain technology. Some blockchain projects have adopted alternative mechanisms, which consume significantly less energy. Initiatives like Ethereum 2.0 aim to reduce the network's energy consumption, and the blockchain community continues to explore ways to develop environmentally sustainable solutions.

Overall, as things stand, blockchain is a complex technology that requires a high level of technical expertise to implement and maintain.

Ethical issues with AI

We need to be aware of the ethical issues of using AI. If at some point humans no longer have the ability to exercise judgement on what is right and wrong, it is not yet evident that it is possible to teach AI to act morally. Tech at present depends on the subjectivity of the person or set of data creating or informing the AI, so will we ever get a truly unbiased AI?

Longer-term, there have been calls for governments to look at regulating the use of AI. People's fears include AI's role in weapons, unseen glitches in systems for financial trading, and fake videos of famous figures or celebrities through the use of voice and image cloning. This increases the potential for frauds and scams.

An example of this is called a deep-fake scam. This is where we have the illusion that a well-known figure is giving advice online, but it is just images of them and their voice that are combined using AI to appear as if it is them. Of course, this is not legitimate.

While we do not yet have precise regulations to protect us in all instances, you can make sure you never click on anything online that you are unsure about, particularly to do with finance. Get the information verified first, as there are many online scams around.

What about if you receive an email from what appears to be the tax office or a recognisable courier company that asks you to *click this link* to receive a refund or delivery? A lot of scammers use these reputable names in order to trick us. Human intuition is a great thing, so always question and think twice, even if you see a message you think is from a reputable source. Unfortunately, these days, scammers are out there in high numbers.

It is good to see these things are being addressed. ComputerWeekly.com reported that the House of Lords has launched an inquiry into the changes that AI will bring about (Saran, 2023). In the article, Baroness Stowell of Beeston says, 'But we need to be clear-eyed about the challenges. We have to investigate the risks in detail and work out how best to address them – without stifling innovation in the process.' The speed of AI development is worth investigation as we require safeguards. It is a positive step to see this in progress.

Regulation of AI

Concerning regulation, the tech industry broadly agrees with there being some form of law to cover AI before things run too far ahead. It will be interesting to see the proposals on how to govern it and if all those who should be involved will be.

The inclusive rollout of any form of regulation needs to bring people and businesses together to have consultations for the public good, as there are many concerns about safety online. I note the UK government is working with its US counterpart, which is a good start. Unfortunately, AI is not always developed with ethics in mind and the information and data gained from AI can sometimes be unethical to publish, especially if it's unverified and used in newsrooms. We need to ensure that ethics take priority in future AI development for the benefit of all of us. This is why a diverse mix of people creating the tech is so important.

AI in talent management

I am interested in the use of AI in recruitment within business. Many companies use AI platforms with predictive analytics to shortlist promising internal candidates, provide tailored career development content, and suggest personalised career paths based on people's goals and interest areas. Recruitment testing using specific software, sometimes called talent management, is being taken up at scale by firms that are swayed by the sales pitches of the software manufacturers.

Companies are increasingly being scrutinised by consumers, employees and job candidates on their standards of corporate behaviour, including environmental, social and governance (ESG) issues. They are being examined when it comes to proving their sustainable and equitable business credentials. Regulators in the UK and US are making human capital reporting mandatory, which could be carried out through a management platform.

This is why talent management systems are increasingly being used. They ensure robust people analytics will be available to companies so they can report accurately and transparently on important ESG criteria, including workforce diversity and gender pay gaps. In other words, a talent management tool helps with the new mandatory reporting requirements.

However, the human factor in organisations, such as conversations with line managers and employees, must occur aside from using the tools. Even computer programs can be made with biases, so it is essential to be mindful of this fact. If the talent management software does not take into account a diverse workforce, it will not be fruitful.

As we learned in an earlier chapter, until 2011, seatbelts were designed with the male anatomy in mind and were tested on males. If a woman was in a car crash, she was 17% more likely to die than a man in the same crash and 47% more likely to be seriously injured. Although this is an extreme example, we need to consider that when they were developed, the seatbelts were being tested on an average male. There was no consideration given to diversity, so when women used them, the seatbelts did not fit properly.

This example shows that we need to increase the diversity in teams right across the business world in technology to reduce the potential for such errors. A broad and diverse workforce always brings better solutions.

This is something we must keep at the forefront of our minds, not only for talent management systems, but for all developments in technology. AI is just one example of where biases could occur. I challenge any programs that are being developed by a workforce that is not diverse to be free of the biases of the individuals inputting the code for these programs.

A bias in this context refers to a pattern where computers can systematically and unfairly discriminate against specific individuals or groups of individuals in favour of others. Although this could be seen as controversial, it is a crucial point to discuss. It is the reason why we must encourage more diversity in developing new tech.

How can we ensure we get around biases in tech tools used in business? A computer doesn't provide a true example of how a human interacts or makes HR decisions, so how can we be sure talent management systems are good for all to use and truly inclusive? Assumptions that may have been coded into talent management systems are something for every business leader to be aware of. This is where more human input is required rather than sole reliance on a tool, and so the awareness by HR professionals is key.

We always need to be cognisant of our own biases, too. Consciously or unconsciously, we all gravitate towards other people who are similar to us because we can relate to them. Think of it this way – bias clearly affects business because a lot of C-suite positions are occupied by middle-class white men, and middle managers tend to be straight white men. In financial services or banking, there are few women at vice president and director levels. Thankfully, more and more companies have policies or guidance in place in recruitment to address diversity in the workplace.

We must all be ethical and fair, especially those of us in senior positions. Biases are a significant area for all leaders to be aware of, as well as their staff. It is fundamental that leaders are constantly reflecting and improving their practice with regards to DE&I.

For those who develop the programs, it is often seen that they are inputting the data based on biases. By this, I mean the programs they develop may expect certain individuals to do better than others or not take account of any differences. When I interviewed Dr Anne-Marie Imafidon MBE, I asked her about this.

INTERVIEW WITH AN EXPERT – DR ANNE-MARIE IMAFIDON MBE

'As more and more company leaders are using talent management systems in recruitment and internally, and these tools use AI, how do they ensure no biases are in the program?'

This is the question I put to keynote speaker, creator of the award-winning social enterprise Stemettes and a recognised thought-leader in the tech space, Dr Anne-Marie Imafidon. Here is what she replied:

'I am a trustee at the Institute for the Future of Work. This is something that we wrote about in a report, a white paper we put together a few years ago called "Machine learning use cases".

'There were three examples in it. One of them, in particular, had a focus on talent management. It was actually a retail company that had written this technology, which we renamed in the user case, that was supposed to help do shift allocations, decide pay and promotions.

'One problem is the data set that your program is pulling in and how that data is collected. As much as we may think data is neutral, it is not, and there are also assumptions when data is collected – the intentions that people have when data is collected (and there are perfect and imperfect ways of collecting information and data). For this one algorithm in particular, the program was collecting data from sources like workplace chat; it was collecting data about where people were on the shop floor. It was also using facial recognition, which should have set alarm bells off. Being able to tell one face apart from the other is one thing, but also being able to read sentiment from people's faces in a way that is scalable and without bias? That was one side of the issue that the company had with the algorithm.

'This algorithm ended up inadvertently doing things like widening pay gaps, including the gender pay gap. The reason for this was the second way the bias came in – namely, the algorithm was optimised in a particular way. Again, the folks that built it did not do this intentionally, or maybe they didn't audit for it during that segment. They had built it to help minimise the pay people were getting and promote them as little and as late as possible without them leaving, so it worked out and calculated how long they'd stay at one level and how likely they were to leave. Promotions and pay rises were offered on a just-in-time basis rather than a way that would have been fairer.

'So, algorithms need looking at from more than just one side. Thought needs to be given to what you are optimising for. Questions need to be asked to establish if what you are optimising for is always fair and right.

'Are you able to optimise on human beings without some sort of bias one way or another? As well as your data collection methods and the datasets that you have, as well as the type of datasets that you are training on – are they also without bias? To answer your question, are talent management systems doing what they need to do? No!

'At the Institute for the Future of Work, we are trying to implement an Algorithmic Accountability Act. That would be a bit like having the data controller and the data processor. On general data protection regulation

(GDPR), you have the person that built the algorithm and the person that is deploying the algorithm, and both of them are held accountable in some ways by law for the impact of the algorithms that end up being used.'

This detailed answer to my question brings up some useful and interesting points. We need to keep a close eye on the outputs from the Future of Work Institute, and it would be great to have the Algorithmic Accountability Act passed. Let's watch this space!

It's clear to see why we need more females in tech! Different views create diversity of thought.

Another question to be aware of is, if we are cognisant of the biases, can we solve them by adjusting AI tools to deliver fairer outcomes for society? Is the issue that people developing AI are unaware of the biases themselves (or that they keep changing)?

The key is that data sets used to train AI systems contain human labels. If the bias is built into the training data – for example, a young man drinking beer potentially categorised as an alcoholic or a drunk – how could this be mitigated? This bias is already embedded in the data sets.

On that note, let's move on to talk about the next topic, or this whole book could soon be taken over by AI!

Online safety

Online safety is not talked about enough. Chatrooms or social sites are places we all need to be mindful of because they may lead to online bullying and harassment.

Since the Covid pandemic, there has been growing online and technology-related violence against women and girls (VAWG). The UN Women, the group working globally for gender equality, has reported that online VAWG takes many forms, such as sexual harassment, stalking, intimate image abuse and misogynistic hate speech.

As so many females have experienced online abuse, this is something being looked into further by the UN Women UK project to strengthen transparency about the nature and extent of online and tech-facilitated VAWG. Because the project is in its infancy at the time of writing, there are currently no reports to share, but these will be good to read once the project is completed.

UN Women UK also runs festivals and music events around the country where the group trains volunteers so that all females are safe when they attend these events. All the events include safe areas women can go to. I have been privileged to apply and be accepted to join UN Women UK and will be taking part in the safety project, to tackle the problems where there are issues with safety for women online. This is a crucial issue to be addressed.

How technology can enhance safety in everyday life

There are some wonderful technological innovations regarding safety both online and offline. An application called Safe & the City provides its users with secure, well-lit and safe routes to get home – just like using Google Maps, but the safety version. Do look this up and share it with as many people as possible to avoid the awful events we hear about in the news, like the murders of Sabina Nessa, Sarah Everard and many others.

Technology and climate change

With the escalating climate crisis, technology emerges as a formidable ally to empower us to tackle the greatest challenge of our time. It offers many innovative solutions to combat climate change and drive sustainable development. From renewable energy advancements to smart grids, carbon capture technologies to precision agriculture, the digital revolution is reshaping our approach to environmental stewardship.

One of the most remarkable contributions of technology lies in the realm of renewable energy. Breakthroughs in solar, wind and hydropower technologies have not only made clean energy more accessible and affordable, but also revolutionised the global energy landscape. The pursuit of efficiency has paved the way for highly effective solar panels, towering wind turbines and innovative energy storage solutions. These all ensure a reliable and continuous supply of sustainable power.

Another remarkable aspect of technology's impact on climate change lies in data analytics and smart systems. The advent of the Internet of Things (IoT) has unlocked a new era of intelligent decision making and resource optimisation. The IoT (also known as smart technologies) consists of devices and technology which connect to the internet and communicate with other devices and technology without any human cooperation.

Smart grids and energy management systems enable real-time monitoring, analysis and control of energy consumption, reducing waste and maximising efficiency. AI-driven algorithms help optimise transportation routes, minimise fuel consumption and enhance logistics, leading to significant reductions in greenhouse gas emissions.

Technology also acts as a catalyst for transformative change across various sectors. In agriculture, precision farming techniques, guided by remote sensing, drones and data analytics, enable targeted resource allocation and the optimising of water usage, reducing fertiliser waste and increasing crop yields. Advanced manufacturing processes, such as 3D printing and circular economy practices, contribute to resource efficiency and waste reduction. Additionally, technology empowers all of us and our communities to become active participants in the fight against climate change through digital platforms, mobile applications and online networks that promote sustainable behaviours, education and collective action.

Beyond tangible solutions, technology fosters a culture of innovation, collaboration and global connectivity. Start-ups and entrepreneurs are driving breakthrough ideas and disruptive technologies, and many multinational corporations are investing in research and development to push the boundaries of what is possible.

As we navigate the complexities of climate change, technology serves as our companion, offering innovative tools, transformative solutions and a glimmer of hope. It is crucial to acknowledge that technology alone cannot solve the climate crisis. It must be coupled with collective action, policy frameworks and a commitment to sustainable practices, but by embracing the power of technology as a catalyst for change and integrating it into our efforts, we forge a path towards a resilient and thriving planet for present and future generations.

This is an exciting area that is only set for growth, which is a good thing if you are planning to work in technology.

Carbon Future Interconnectors

In an era where the battle against climate change is at the forefront of global agendas, technological innovation has emerged as a beacon of hope for helping us towards a sustainable future. There is a ground-breaking development called Carbon Future Interconnectors, which consists of an interconnected network of energy transmission systems. They have been designed to revolutionise our approach to carbon reduction.

These cutting-edge interconnectors are like the vital arteries of a cleaner, greener world. They connect renewable energy sources, storage facilities and consumers across vast distances. Interconnectors enable the efficient transport of renewable electricity, effectively bridging gaps and unlocking the potential of remote regions to be rich in clean energy resources.

The significance of Carbon Future Interconnectors is not only in their ability to transmit electricity, but also in their capacity to address the intermittent nature of renewable energy sources. For example, by interconnecting various renewable energy generation sites, including wind farms, solar power plants and hydroelectric facilities, all these networks establish a dynamic and balanced ecosystem of clean power. Also, with the integration of advanced energy storage systems, excess energy generated during peak periods can be stored and redistributed during low-generation periods, ensuring a constant and reliable power supply.

Beyond any immediate benefits of reduced greenhouse gas emissions and improved energy security, Carbon Future Interconnectors hold immense potential for global collaboration and shared progress. They can bridge gaps between nations and continents, facilitating the exchange of clean energy across borders and fostering cooperative efforts in combating climate change. By going across geopolitical boundaries, these interconnected energy networks allow nations to align their climate goals and forge alliances centred around sustainability.

In addition to their transformative impact on the energy sector, Carbon Future Interconnectors also catalyse economic growth and job creation, which is good news. As the demand for renewable energy increases, the interconnectors serve as catalysts for investment in renewable energy projects, driving innovation and propelling the growth of green industries.

It is evident that technology transcends geographical borders, inspiring international collaboration and the sharing of resources, knowledge and solutions, ultimately accelerating progress towards a sustainable future.

There is so much that I could have covered in this chapter. AI alone could fill a book, but I have chosen what I consider to be the most topical areas for you and given you an overview of their current status.

You have seen a few examples in the vast array of the latest technology, from AI to crypto and blockchain to online and offline safety, and even how tech can help in the fight against climate change. I recommend you pick at least one emerging technology to research further.

The potential of blockchain and crypto to change many things for us, including the way we bank, and the fact they help those who are unable to reach financial centres in the way that we take for granted is a positive. However, please research and learn all about crypto before you invest or take part.

In the next chapter, I highlight some of the common job roles you will find in tech. We will hear about the experiences of women who share their insights into some of the roles the chapter will cover.

FOUR

Tech Trade: Unveiling Common Job Roles

> 'Life was simple before World War II. After that, we had systems.'
>
> —Grace Hopper

Women in tech have come a long way, but there is still much work to be done. We all need to keep thinking about all the opportunities that exist.

Before we delve into the details of some of the common roles in tech, I would like to take a moment to thank the women who paved the way for us. Here are the names of some of the top women who changed the tech world forever (Kraus, 2018; Wyman, no date):

- Ada Lovelace: the world's first computer programmer

- Carol Shaw: the first female video game designer
- Elizabeth Feinler: designed the original search engine
- Grace Hopper: an esteemed computer scientist
- Hedy Lamarr: an inventor of wi-fi
- Joan Clarke: an Enigma codebreaker (Bletchley Park)
- Katherine Johnson: NASA mathematician
- Karen Spärck-Jones: a pioneer in information science
- Radia Perlman: the mother of the internet
- Susan Kare: a graphic designer behind Apple's iconic graphics

You may or may not have heard the names of the women presented here. Feel free to look up more information about them as each woman has an interesting story worth reading. Also, do Google some more names; there are many unsung women in tech. This will enable you to see the potential as these are 'hidden' figures. We require more females in tech to be able to celebrate our contribution. Let's commit to creating a more equitable future for women in this industry. After all, creating a big change is about making a small change every day.

In this chapter, we will look at the many aspects of general roles in tech. We'll look at what they do and some essential skills they require. You may notice many of these roles require soft communication and interpersonal skills. While I recognise that developing certain technical skills is essential, some roles are more business and customer-focused, so the types of technical skills you may need can vary greatly.

The chapter will only cover the common roles in tech, the ones that you will tend to hear about in the industry. This is because the more specific roles are forever changing as new areas of tech come out regularly, and some roles take on fancy new titles, so I have focused on describing the evergreen or classic types of roles where possible. These roles are found either in projects and programs or generally in technology, and the chapter will help give you a flavour of what you might enjoy. I will include my interviews with women in some of the roles I mention so that you get to hear what the work is like from a practical and day-to-day aspect.

Before you go into a tech career, a basic understanding of IT and computing is good. There are plenty of ways to gain this knowledge, such as short overview courses, community groups, boot camps, internet research, public library research and speaking to people already in the tech sector. Some positions require more advanced skills and a better understanding of computers and IT than others. There are many resources available online if you need them, and I share some in the resources section.

Another reason to make sure you have a basic understanding of IT is that it will help should you decide to pursue another role within tech further down the line. Interests change over time as you move into a new sector, and as we've seen, career journeys never go in a straight line. As the tech industry spans many different areas, it can be beneficial to have a good grounding in the sector you are going into.

Technology drives innovation across industries, so having knowledge about the common roles in tech will open doors to opportunities. It allows you to make informed decisions about a potential career path, enabling you to align your skills and interests with the demands of the job market, which will benefit you. I have worked within projects in many different industries and the same principles apply to my job role, no matter where I go. Building up your knowledge on what is out there assists as you make any change.

I share information on these roles to give you a holistic perspective on what you will find in all areas within the technology sector, plus some of the variations in names to help you recognise them. With some insight into these roles, you can confidently navigate and identify areas for growth to fulfil your goals in the ever-expanding area of technology.

Project manager

Some related roles and common job titles include project coordinator, project support assistant, project leader, work-stream lead and project analyst. Most of these roles tend to act as more of a support to a senior project manager or programme manager.

If you are interested in becoming a project manager, the first step could be to gain experience either in one of the roles listed above or in an analyst role, for example. This would give you some skills and grounding in subject areas in the business, and in some companies you can transition after gaining experience from this role into becoming a project manager. This is not true in all companies, though, so do research potential employers as well as roles.

What they do

As the name suggests, a project manager is responsible for the successful delivery of a project. A project is a temporary activity or series of activities that ends in giving a specific result. For example, your project could be to deliver a new product or service, implement a software solution or improve an existing process in the business.

The important thing to remember here is that a project is *temporary*, so as a project manager, once you deliver the requested product, service or improvement, you are likely to be straight on to the next one. You may even be running more than one project at a time.

All projects have a start date and an end date for when you need to have achieved the required result. In addition, all projects have constraints such as funding, time or resources like people and specialist skills that you may call upon to help in delivering the outcomes of the project. When the resources are people, they may be assisting with their knowledge and expertise in addition to working in their day jobs, so you as the project manager have to manage their time through influence and negotiation.

Did you know project management actually originated from the building and construction world? Effective coordination and management of construction projects has always been essential to ensure successful outcomes, so with regard to early project management practices, it would be fair to say that the evolution of the role came from ancient times, such as the building of medieval cathedrals, Roman aqueducts and even the Egyptian pyramids. All these required intricate planning and a lot of organisation.

This is what the building trade does as everything is meticulously and logically planned, for example, digging foundations, sourcing all the required materials and having the resources (human power) in place before the work starts. Project planning is an essential part of this to ensure a successful build.

The term *project management* would not, of course, have existed in ancient times; yet all the same principles applied then as now, such as timelines, materials, labour and resources. The discipline became formal in the twentieth century through engineering and defence industries. The project manager role varies depending on the industry and / or the type of project.

At the start of a project, the project manager ensures the goals are clear and will meet with stakeholders (those interested or involved in the project delivery) who can be external or internal to the organisation. They could also be in different areas of the company or business. The project manager then determines what can be accomplished with the budget, timelines and resources available. They will create a detailed project plan and gain agreement on it from the various stakeholders (which is important for success).

The plan includes key milestones and deliverables, and the project manager ensures throughout the project it is kept up to date. They hold regular meetings with the owner of the project (the team or person requiring the delivery of the product or service), which can be set by the owner, to share progress and discuss any problems or blockers. These meetings can be weekly or monthly, depending on the length of the project and as the meeting progress is shared, any problems or blockers can be discussed and, if required, escalated.

The project manager also maintains a risk and issues log. It is their responsibility to maintain and monitor risk management, which is an essential activity within the role, so that they can flag anything that could cause the project to be delayed or go over budget, or where items are out of the scope of the original delivery, or any other blockers preventing the project's success.

One common blocker is a stakeholder being prevented from completing a task that's essential to the project's success. In that case, the project manager would intervene, and if there are conflicts with stakeholders, communication skills are vital, as are negotiation skills.

Time, cost and scope are always at the forefront of the project manager's mind as one affects the others. If a project gets delayed, it may mean more cost, or if the scope of project is increased, the project manager may need the resource of people for longer.

The lifecycle of a project does, of course, depend on the industry you are working in as a project manager. In tech, software projects often follow a specific software development lifecycle.

Key skills

Project management is a people job, so navigating different personalities and being able to make sure they all get the job done is paramount. In this role, essential areas are:

- **Communication**: Effective communication with all stakeholders is essential to ensure they are informed and foster collaboration.
- **Scope**: This means defining the project's objectives, deliverables and any constraints. The project manager needs to clarify the project's scope at the beginning.
- **Cost**: This is estimating and controlling project costs, for example for materials or resources such as people and equipment, and a contingency fund. Monitoring the budgets is a regular activity for a project manager.
- **Time management**: Realistic schedules need to be created and managed for timely completion of a project. It is key for a project's success for it to be completed on time.
- **Quality management**: Implement quality control to ensure standards are met or exceeded. On a new software system, for example, quality would be a high priority.
- **Risk management**: Identify and assess risks, and reduce or remove them before they impact the project's success.
- **Stakeholder management**: Identify all stakeholders and engage with them, address any concerns they may have and manage their expectations.

If you have watched *The Apprentice* on TV, you may have noticed how candidates can forget the people side of their management skills (even forgetting to communicate with their own teams sometimes), and focus all their effort on the delivery of a product or service or the number of sales in order to win at a task. This often has the opposite effect, losing them the task because they have not managed the people around them in their teams well enough to ensure their overall success.

A project manager is fully responsible for following methodologies to help with the smooth delivery of the project and bring it to a close. Common methodologies are Prince2 in the UK or Project Management Professional (PMP), which is a globally recognised certification. Sometimes, organisations have their own adopted methodologies, which are based upon similar principles and aligned to the way the business operates. A methodology helps to ensure no additional work or rework is required once the project is completed and it can be handed over to a business-as-usual team. Upon completion of a project, the project manager usually writes a closure report, including any lessons learned.

On *The Apprentice,* candidates are often nominated or nominate themselves to become the project manager, which gives the role a somewhat rosy view of being in the limelight – apart from when the candidates fail the tasks and endure Sir Alan Sugar's boardroom grilling. They tend to fail because they do not realise the extent of the work being a project manager requires, particularly management and dealing well with people to ensure they complete the work.

This is why I put communication first at the top of the list of skills, as it is so important to articulate and share information with all stakeholders quickly and clearly. Sometimes in tech, there are areas where stakeholders in the business may not understand the jargon. It would be your job as project manager to communicate it in a way they can take it in and appreciate.

Everyone in an organisation will have different priorities, and project managers need to work with many people and stakeholders, many different personalities and sometimes external vendors, too. For example, a project manager may have to ensure external parties complete the work on time for a part of the project. Negotiation skills are therefore a must to arrive at agreements to deliver a solution that all can work with.

Being organised and able to plan is another important skill for a project manager. You will be continuously preparing and ensuring all stakeholders understand the plans, and you will need to keep them up to date. Time management is key as the project's success is dependent on it being completed and delivered on time.

Business analyst

This role focuses on identifying business issues within a process, product or service and using technology to address them. Typical job titles include technical business analyst, functional analyst, business systems analyst, business process analyst, IT business analyst, project coordinator, analyst, computer systems analyst, systems analyst and user experience analyst.

Business analysis is about understanding the goals of a business, identifying any challenges that may be affecting progress, and then finding appropriate solutions to address them. This could be within a particular area or function in the business, or even the business as a whole.

Business analysts may be asked to help improve or change an existing process within the organisation as a whole or a particular department that may not be serving that part of the business with value or benefits any longer. They do this using technology, and the work required could perhaps be choosing a new software system or even helping with the building of a bespoke software application. Technical business analysts balance the needs of the business, the users and what is technically possible, so they require technical skills as well as a good understanding of their business and industry.

If you become a technical business analyst, you may work in an IT department or software development project team, and you could be part of a small or large delivery team. As a business analyst who is not involved in technical improvements, you may work as part of a project team to improve a service, process or product. In this role, you would be required to look in depth at and analyse the current process or state of the product or service before making any recommendations on the improvements.

What they do

Business analysts spend a lot of time asking questions to understand what challenges exist and what the desired end result is from the area requiring improvements or even a new system. They speak and interact with many stakeholders (this can be both within and outside the organisation, and at different levels) who will be impacted by the end result or delivery of a project. These interactions can be through one-to-one interviews, group brainstorming sessions, the setting up and use of focus groups or conducting surveys.

The types of questions business analysts ask of stakeholders include:

How are they working with an existing process or service?

What are the problems or challenges they experience with it?

Why would they need it to be fixed or changed from the current way of working? This process is referred to as *requirements gathering*.

Business analysts also spend a lot of time documenting and analysing the data they receive. They can be responsible for any of the following:

- **A business requirements list**, sometimes called a business requirements document (BRD): Using responses from the requirements gathering, this document details the features, fixes and / or new process that must be in the final version of the proposed solution.
- **User cases**: These documents detail the exact steps or actions that a stakeholder needs to take to complete a task or achieve a goal using the new process.
- **Diagrams**: A business analyst often produces diagrams rather than long, wordy documents to describe a process, and they will use different diagramming techniques to show the current process. This makes it easier for anyone to see and identify the parts of a process that should be fixed or removed. A special software called Visio is good for flow diagrams and is often used to show process flows and / or mapping by the analyst.

When the business analyst has completed their analysis, they report back to stakeholders with a recommendation for the right actions that should be taken. Depending on the business and the issue being addressed, plus the priority of the problem to the company and timing, the correct course of action can vary. The factors to consider, of course, also include budgets and funding to carry out the work.

Business analysts may be expected to implement or help carry out the proposed solution, which again depends on the organisation or set up of the project. If they do help carry it out, they are usually part of a project team and work closely with the project manager. For smaller pieces of work, business analysts often deliver the actual project. This varies based on the size of the work and the way the company or function in the structure of the business is set up.

Actual project delivery will be done by more experienced analysts. I recommend, if you are interested in projects, that a support or analyst role is an excellent route to learning more subject matter with the potential to progress.

A business analyst may also be tasked to help with programming, data analysis, project management or project coordination, and/or training duties once the new solution is ready. As this role is so close to the detail of the proposed change, it makes sense for a business analyst to be involved in the training, whether it is in the use of a new system, enhancements to an existing system or changes to a process.

Key skills

Communicating and interpersonal skills – being able to build and cultivate relationships with stakeholders are crucial in most roles, but even more so for the business analyst role where you need to ask them for information. Importantly, stakeholders who like and trust you will be more open to your requests and offer helpful information to you. I will cover these skills in more detail in a future chapter about networking and personal branding.

Analysts need to speak to, interact with and interview many different stakeholders, such as business leaders, senior managers, end users (of the proposed change or new solution) and technical teams. These are all people with diverse backgrounds, experiences and ways of thinking, so the business analyst needs to be clear in their questions and requests to ensure their messages are conveyed accurately and well, and that they receive the correct information in response. There is a need to drill down into the problems for analysis. Being able to create good working relationships and understand their stakeholders while articulately presenting ideas and solutions is essential for the business analyst.

It is key for analysts to be able to ask great questions so they understand stakeholders' needs or requirements. This is crucial to draw out the correct information to address the problems and recommend a solution. Analysts also need to be prepared to present and sometimes even defend their recommendations to stakeholders and their teams at meetings for the best outcomes of the work.

A lot of the business analyst's responsibility lies in documenting information, so for this role, you need to be comfortable with business writing and presenting the information in visual form. Business analysts review and summarise quantitative and qualitative information (which can be a lot of data), and they need to be able to suggest what they think is the best way forward from their analysis of the data, so this role requires critical thinking and problem-solving abilities. A business analyst will manipulate and extract data, often using Excel (again, the tools they use are dependent upon the organisation or business). If they are technical, they may deal with databases to manipulate the required information (this is all dependent on the nature of the original problem).

Analysts often create different types of visual diagrams and models or mock-ups, such as flowcharts to represent the steps involved in a work process. They also represent how data flows in and out of a process or business system through what are, appropriately, called 'data flows' in the diagrams.

A working knowledge of how to use diagramming software, such as Microsoft Visio, is helpful if becoming a business analyst appeals to you. This is a software commonly used in business, but there are many others that show flowcharts and relationships between how processes fit together when doing analysis. There may be software built in-house in the company you join, but the principles of use will be the same.

Key entry and educational requirements

There is no specific set of qualifications or educational requirements to become an analyst. Many technical business analysts have a computer-related undergraduate degree, but some join a company simply with A-Levels or equivalent and a knowledge of the industry. There are even analysts with degrees in geography, art or other apparently unrelated fields. You can always receive technical training through additional courses and on the job.

I know people who started in different roles and trained on the job to become a business analyst. You can receive training in most roles you are keen to pursue, although this is dependent on the needs of your employer. It is beneficial to job shadow more experienced people than you as part of your development plan.

There are, however, certain skills, knowledge bases and qualifications commonly sought by employers looking to recruit in this field. If you are thinking of professional certifications or a degree, then business-related fields such as business systems, information systems, business administration, economics or finance can provide a solid foundation for a career as a business analyst. Other further education routes into a business analyst role include:

- British Computer Society (BCS) Foundation Certificate in Business Analysis
- BCS Certificate in Business Analysis Practice
- BCS Certificate in Modelling Business Processes
- The International Institute of Business Analysis Certified Business Analysis Professional (CBAP), which is recognised globally
- Certification of Competency in Business Analysis (CCBA), which is recognised globally

For additional information on certifications related to this role, please refer to the BCS website. You will find the address in the 'Resources' page.

Now let's read what a senior analyst says about what she does in her role. As described above, this technical business analyst works in the context of an IT department, specifically in IT infrastructure. In this interview, she highlights that self-belief and persistence paid off for her in securing her role as an analyst in tech. For some important points, first-hand experience and sound advice, please read on.

INTERVIEW WITH A PROFESSIONAL – AKUA OPONG

Akua Opong is a senior analyst in service management and a chartered IT professional with over six years of experience in financial services. She is currently at the London Stock Exchange Group (LSEG).

Here is my question and answer session with Akua.

What do you do?

'I work in corporate technology, supporting colleagues with their hardware and applications, and providing second-line support on a daily basis. I am process-driven and enjoy creating technical documentation, and have a methodical and analytical approach to problem solving while maintaining the highest of standards. I am passionate about technology, with the ability to learn new technology and adapt to different situations.

'One of my strengths is to work efficiently to build knowledge of systems, hardware and software. Also, I have acquired a broad knowledge and understanding of all aspects of computing, networking and troubleshooting skills.'

What do you love about the work as a senior analyst?

'My role is client-facing; this allows me to meet new people across the business, collaborate and resolve their technical issues. I learn about new technology daily and the most effective way to use it. Every day there are enhancements in technology, so this role involves finding out about these advancements, attending webinars, networking and reading articles.

'Knowledge is power. You need to be team-orientated, collaborative and analytical, enjoy reacting to changes and follow the IT Infrastructure Library (ITIL), a collection of volumes describing a framework of best practices for delivering IT services.'

What inspired you to become a senior analyst?

'In my younger years, I wanted to be a paediatrician – a medical practitioner specialising in children's healthcare. In my late teens, I thought about working in the army in emergency response and IT intelligence roles. I didn't end up working in either of these careers, but my passion for helping others naturally progressed me into technology, where I have worked for over eight wonderful years supporting thousands of people globally who have access to tech, providing an act of service.'

How did you become a senior analyst?

'After completing my studies, I encountered several challenges in securing my first role within the IT industry. However, I persevered and took on a position in retail while continuing my search for opportunities. Eventually, I secured a role as an IT support analyst, which enabled me to gain experience in a variety of industries, such as hospitality, consulting and financial services.

'In July 2019, I had the opportunity to join LSEG as a senior associate. The fast-paced environment that combines technology and finance has been an exciting and fulfilling experience.

'I studied business with sociology and religious studies/philosophy, followed by a BSc (Honours) in computing and IT (sandwich course) at the University of Surrey.

For my dissertation project, I developed a smart key concept and completed an excellent work placement with infrastructure support for Rolls-Royce Motor Cars Ltd.

'My first IT job was supporting the Dubai royal family across their UK properties. I worked at Rathbones for nearly five years, then PA Consulting. Fast forward to July 2019, when I joined LSEG to work in the Europe, Middle East and Africa desktop services team within corporate technology. It is very fast-paced, which I love.

'Representation is really important to me as an individual, as is being part of a campaign or project that brings an idea to enhance diversity to reality and achieves a positive outcome. There have been challenges and difficult moments, but through continuous learning and the work that I have been doing in charity, wellbeing and DE&I, I know I can help others to thrive and add value.

'Some of my current skills include being a communicator, connector, collaborator and showing courage. I like to lead by example and give others an opportunity, so I am interested in education for women, gender equality and eradicating poverty.'

What is your typical day?

'Each day is different. I like to plan my day using the Pomodoro technique, where I focus on the hard tasks first, and then work on the simple tasks. My tasks could include providing support to new starters with their IT setup, assisting a user with application queries, onboarding employees, IT procurement, early careers induction and an IT sustainability project. I sometimes work on the "tech bar" (this is a physical help area), resolving technical queries, doing mobile device setup, configuring IT hardware such as laptops or responding to a "Help! How do I...?" query. My key responsibility is providing second-line support as a point of escalation for queries that cannot be resolved by the service desk team.'

What advice would you give to other women interested in pursuing a similar role or career to yours in the tech industry?

'Be open to multiple possibilities, but don't overstretch yourself or allow imposter syndrome or a lack of confidence to stop you. Focus on the here and now, what is important to you at this moment. Be true to what it is that you want to accomplish – your goal is key, so be brave enough to go for it!

'This includes looking for role models in the industry, those helping to set a benchmark for others. When we see others that look like us in a room, it gives us a chance and makes us feel like we belong. Never stop believing in what you can become in the future. There are times in life when you hit a stumbling block, but you need to keep going and just persevere.

'I attended Michelle Obama's *Becoming* book tour at the O2 in London. She is a massive advocate for young girls to be given a chance to succeed, regardless of their background. Each person should be given opportunities in a world that is fair and equal. It is about uplifting, empowering; not about competing with each other, but working for each other – for the girl or woman's voice to be heard.'

Anything else you would like to share?

'I recommend the film *Hidden Figures* as this focuses on three trailblazing women in STEM. One in particular, Katherine Johnson, a mathematician who once worked for NASA, said, "I didn't do anything alone but try to go to the root of the question - and succeeded there".'

Consulting

Some common job titles include technical consultant, IT consultant, technology consultant, management consultant. Consulting can be similar to business analysis and there is sometimes an overlap between the two.

At a high level, people in both roles need to understand a challenge that a business is facing and guide it towards reaching its preferred outcome. However, technology consulting focuses more on a strategy or direction that a business needs to take rather than concentrating on an issue that may require more specialised analytic expertise, so it is good to be aware of the difference between the two. Consultants in traditional corporate businesses would be more senior than analysts with greater experience and skillsets as the role is often more strategic.

If a business leader is thinking of a large alteration to their IT infrastructure, the consultant would need to understand what the rationale is for the decision, what is causing the business leader to consider this change. The change could be renewing software across the whole company, or perhaps the business putting some or all of its applications and data on the cloud and using a service provider to manage this, rather than taking up a lot of space to accommodate all the physical computers and hardware required for this tech. If space is an issue, using the cloud would be a far more efficient solution, so a technology consultant would come in and look at the business leaders' reasons for making the change.

A consultant is like a trusted advisor for a business project team, brought in for their knowledge and experience. Among many other things, they want to know what the timeframe for making the change is and the budget allocated for the work.

Any advice they give and recommendations they make are neutral, which means for example that any solution they propose, such as new software, would be what they deem to be the best option for the business's requirements. It is not a one-size-fits-all solution that is favoured by the consultant. Even if they do happen to have ties or contacts with a particular solution provider, they still have to remain neutral in all their recommendations.

You will find many consulting firms that offer a broad range of services to clients. The smaller or 'boutique' consulting companies have fewer than 100 employees and may focus more on local rather than international clients. From a technology perspective, consulting firms tend to fall into two areas: IT strategy, meaning helping companies create a strategic roadmap from where they currently are from a technical perspective to where they would like to be, and IT operations, which help companies understand where they can make improvements in their infrastructure and systems, and then implement the proposed solutions.

What they do

The life of a consultant varies greatly. Their work depends on what the client asks them for per the statement of work agreed between the client and the consultant/consulting firm, and no two projects are similar.

Consultants' work can involve a lot of travel to client sites, which may be abroad. A project can last from weeks to years. A lot of time spent travelling and working long hours is not for everyone, so if you seek normal office hours, you need to take this into consideration if you fancy the idea of being a consultant, especially if you have other commitments.

Key skills

As a consultant, you will be expected to chair and organise a lot of meetings with clients and teams, so you need to be able to present information to any type of audience. Rather like the project manager, you will have many stakeholders to consider and may even have a team working for you or with you, dependent on the size of the project.

Being able to communicate well is key to overcome any differences of opinion and deliver messages in a respectful and clear way to everyone involved. You will also need excellent presentation skills. Building good business relationships internally and externally will be important to the project's progress. Consultants often work in small teams that contain members from both the consulting firm and the client's organisation, so interpersonal skills are essential.

Like the business analyst, consultants review a lot of quantitative and qualitative information, so critical thinking and problem-solving skills are a must. You need to be able to summarise your recommendation articulately and find the best solutions, and then show the client the way forward. Business writing is also a core part of the role, which includes a lot of reports and presentations.

Consultants are often considered to be subject-matter experts in their field by their clients. Being a subject-matter expert requires you to keep up to date with any developments and advancements, and typically stay in the know. This desire for continuous learning is common throughout the tech industry.

It is possible for a consultant to be put on a project where they have little or no experience with the client requirements. In this instance, you would be expected to get up to speed quickly through self-study or quizzing people in the teams who are experts in the subject. Finally, you will often be working within fixed timeframes, so managing your time and minimising disruptions is crucial.

Key entry and educational requirements

Consultants will need to have both business and technical expertise. You will find many entry-level consulting internships and jobs with major firms like Deloitte, Price Waterhouse Coopers, KPMG and Accenture, as well as smaller consultancy firms. Usually, these firms require you to have an undergraduate degree, but consultants often hold degrees in a wide variety of subject areas that are non-business related.

Speaking to people when you are out networking is always a good way to find out more about entry into this or any field. Alternatively, you could look at a consulting firm's website. This is often a good place to start your research as it helps you understand the company strategy and values, and what the company looks for in candidates, along with any entry-level opportunities and detailed information on recruiting practices. Candidates with two or more years of professional experience after college or university are recruited all year round.

Software engineer

Common job titles include software developer, programmer, coder and software architect. A software engineer brings a tech application, which we may use on our phone or laptop, to life. Their role is to design, develop, test and maintain software applications and systems. We are always reliant on tech applications, so good software engineers or developers are constantly in demand.

The terms coder, programmer and developer tend to get used interchangeably, but they can have slightly different meanings depending on the context. A developer has broader skillsets and participates in the entire software development lifecycle. They are involved in various stages and have a holistic approach to software development that encompasses both coder and programmer skills. A coder focuses on writing the code and the basics of programming with the rules of programming languages and may, for example, write part of the program. A programmer will write, test and maintain computer programs and can create software solutions from scratch, requiring more in-depth thinking and analysis. There can, of course, be an overlap in the responsibilities and skills of the three roles.

What they do

A software engineer will analyse the end user's requirements for the solution, create technical specifications, and then write the code using a programming language and frameworks. They collaborate with teams, including software developers and project managers, to ensure the solution they provide meets quality standards and best practices and, of course, that they deliver on time. They often participate in troubleshooting problems and debugging software, and they optimise software performance to make sure it is efficient and reliable.

They will write the code first, and then test it and make modifications. They create documentation throughout the software development process to serve as a reference to assist with maintenance of the code or if there are fails in the software. Sometimes, they assist with training end users on how to use the program.

There are terms that are regularly used in the world of a software engineer: front end, back end and full stack. It is good to know the difference.

The front end means the software engineer focuses on the part of the program the end user, like you or I, will use, so they are primarily interested in the actual user experience. The back end refers to the application where all the data is stored – the database and all the logic. This is really the centre of the application, not the back end. The full stack, as the name implies, means the front end and back end.

Key entry and educational requirements

You usually need an undergraduate degree to become a software engineer, but these days, there is so much demand for this role, you may be able to learn by yourself through online courses, boot camps or self-study. There are also online groups (see 'Resources' and my upcoming interview with Asia Sharif) that you can join. However, a computer science degree offers a more comprehensive grounding as it covers the architecture of the computer, different operating systems and networking so you become familiar with and understand the hardware and how things all link together, which I would say is more of a technical role.

INTERVIEW WITH A PROFESSIONAL – ASIA SHARIF

Asia is a multi-award-winning software engineer, public speaker, blogger, poet, engineering mentor and business founder. It is useful to hear from someone working in the role of a software engineer, so I am keen to share my interview with Asia here to give some insight to you.

What inspired you to become a software engineer?

'To embark on this journey, I primarily needed a love for learning. Being self-taught, I have had to constantly seek out resources, online courses and tutorials to acquire the necessary programming skills. The ability to adapt and grow is crucial as technology evolves rapidly, requiring continuous learning and staying up-to-date with the latest tools and languages.

'Additionally, problem-solving skills have played a vital role in my path as a software engineer. Analysing complex issues, breaking them down into manageable components and finding effective solutions are essential skills to possess. Collaboration and communication skills are also important in working effectively with diverse teams and translating technical concepts into understandable terms for non-technical stakeholders.

'Ultimately, my inspiration to become a software engineer stemmed from a desire to challenge norms, inspire others and make a positive impact in a field that has historically lacked diversity.'

What do you do day to day?

'Day to day as a data/software engineer, my responsibilities revolve around working with data and building robust data processes. This involves tasks such as data extraction, transformation, loading, datamodelling and creating data pipelines. I collaborate closely with cross-functional teams, including data scientists and analysts, to ensure data integrity and to optimise data workflows.

'What I particularly enjoy about this role is the problem-solving aspect. Each day presents new challenges, which keep me engaged and continuously learning. I find great satisfaction in overcoming technical hurdles and finding efficient and innovative solutions. It's a role that allows for personal and professional growth as the technology landscape evolves rapidly, providing opportunities to explore emerging tools and expand my skillset.

'As a data/software engineer, I need several essential skills and qualities that contribute to success in the role. Proficiency in at least one programming language is crucial, as it forms the foundation for developing software and data processes. Problem-solving skills are equally important, enabling engineers to analyse complex challenges, break them down into manageable components and devise effective solutions. Additionally, a curious mindset plays a significant role, as it drives continuous learning and exploration of new technologies and techniques.'

What advice can you give about your role or career in the tech industry?

'My top tip for women looking to start a career in tech is to throw yourself in at the deep end. Don't be afraid to take on challenging projects or opportunities that push you outside your comfort zone. Embrace the mindset of continuous learning and growth, as the tech industry is constantly evolving.

'Another valuable tip is to become a mentor. As you progress in your career, share your knowledge and experiences with others who are just starting out. Being a mentor not only helps others, but also enhances your own skills and reinforces your expertise. It's a fulfilling way to give back to the community and contribute to the development of future talent in the industry.

'Attend conferences, workshops and meetups where you can connect with like-minded individuals, exchange ideas and build meaningful relationships.'

Can you tell me what you think about AI?

'Simply, it is a great addition to the future of technology in every aspect.'

Next, I want to share an interview with Kasia Wojciechowska about her career in tech, showing how her role differs to Asia's in that she is more client facing as a non-engineer. Kasia has seventeen years of experience in the recruit, train and deploy arena, which is known for bridging the gap between academia and the workplace by creating the next generation of IT and business professionals. She has achieved many things, and has exciting nuggets of success and advice to share.

INTERVIEW WITH A PROFESSIONAL – KASIA WOJCIECHOWSKA

Tell me a little about what you do.

'Currently, I lead client activities in the financial services and insurance sector, partnering with clients to build their future talent pipelines via tech boot-camp programmes. In my role, I create and maintain strong working relationships with clients and partners, leading on pitches/proposals, leveraging sector expertise and insight, and owning the client value proposition. I collaborate with various internal and external stakeholders as well as attending industry events.

'Internally to the company I work for, I focus on inspiring and giving direction to junior team members. I am part of a mini-management team responsible for driving the commercial success of our programme.'

What skillset does your role require?

'This role requires strong interpersonal skills and the ability to communicate effectively with a variety of stakeholders. It also requires a strategic and consultative mindset, experience working with global clients, and creative and innovative thinking in developing new initiatives.

'I enjoy the variety of stakeholders that I liaise with, as well as the opportunity to help to build the next generation of future tech professionals. I mentor junior tech and business professionals, including women in tech.'

What is your advice on the key skills and qualities required to start a career in tech?

- **Intellectual curiosity** – stay curious, seek new knowledge and expand your skills throughout your career.
- **Networking and relationships** – invest time in building connections with colleagues, mentors and industry professionals.
- **Stretch goals** – be willing to take on challenging roles and responsibilities, and take calculated, measured risks outside of your comfort zone.
- **Core path** – pick a core skill that aligns with your natural abilities, eg, analytical skills, mathematical skills or creative/visual skills. Build a solid foundation around this, but add other transferable skills into the mix.

'To sum up,' says Kasia, 'find joy and purpose in anything you do. Learning to code or working with data can help prepare you for a transition into a number of different roles in a digital world. Remain focused, adapt to changing circumstances and always maintain a positive attitude.'

Kasia provides a great overview with tips you can use as you progress along your journey.

There are some key roles I have not mentioned yet in this chapter. In today's world with the large amounts of data generated, these roles are set to grow exponentially. One example is data scientist. The role of a data scientist is to collect and analyse complex data from a range of different sources, using machine learning tools, AI, data mining and statistical tools. As it is such an involved role, it usually requires candidates to have an undergraduate computer science degree.

Another related role is data analyst. While this does not require a degree, it still has the benefit of being analytical. As we will see from this interview with someone already working in the role, analytical skills and a love for data and statistics are key.

INTERVIEW WITH A PROFESSIONAL – DATA ANALYST

I asked a woman with a background in HR what inspired her to become a data analyst. This is what she told me:

'I decided to take a career break and focus on a new career journey, so I registered for online data analyst training. From there, I realised that data is beautiful, data is everything, and we can all use it to our advantage. For example, an organisation can use data to find out where it should channel its advertisements and cut costs on any streams that are not bringing in a lot of clients.

'I recommend joining the tech industry, because it does not matter what your set of skills is. I started off in HR, but I found it so easy to change my career into data analysis.

'As a data analyst, you need a combination of technical skills, analytical abilities and personal qualities to be successful in the role. For example, a data analyst needs to have strong Excel skills for data manipulation, calculations and basic analysis. Additionally, they need to be able to create clear and insightful visualisations using tools like Tableau, Power BI or Matplotlib to present findings and patterns effectively. It is important for a data analyst to have good communication skills to be able to explain their complex findings in a clear and compelling manner to both technical and non-technical audiences. In addition to this, a data analyst needs to be able to pay attention to detail, being meticulous in data cleaning and validation to ensure accurate analysis and reliable results.

'My day to day starts with gathering relevant data from various sources, which could include databases, spreadsheets or other data repositories. Once the data is ready, I perform exploratory analysis to identify patterns, trends and outliers in the data.'

This interview gives us a brief insight into what the data analyst role means and tells us that it is possible to transition from a completely different background and use the skills and experience from a previous career to make this transition successful.

Cyber or information security is another rapidly growing area. Although it is good to have a degree for foundation knowledge, once again, you can join this area through other disciplines or self-study. There are many professional development opportunities as cybersecurity is and looks set to remain very much in demand.

For the last interview piece in this chapter, I asked Dr Anne-Marie Imafidon MBE to share her top advice as a guide to someone considering a career in tech.

INTERVIEW WITH AN EXPERT – DR ANNE-MARIE IMAFIDON MBE

'There are lots of roles, and they change so quickly and sometimes become new roles. For example, a product manager's career progression is not always clear.

'The whole thing is about transition; the roles do not matter. Not everyone should be a designer first or in development first. For some people, they can enter as a developer; for others, data could be an entry route. I would recommend giving everything a taste, like you would at a buffet, before you decide what is for you.'

I love this analogy from Dr Imafidon, and it answers my request in a nutshell. Who doesn't like to try everything at a buffet?

It is what I did. The MSc degree course gave me a lot in that year of study – it provided me with an all-round view of what was available to me in tech. I disliked programming and instantly knew it was not for me. This is why it is valuable to gain practical experience when you learn or take courses.

Throughout this chapter, we have looked at the variety of tech roles and spoken to real people who have shared with you about their day to day and the skills they learned or acquired when starting out. We have seen that there are many roles and different ways to enter those roles in the industry.

Use your strengths and find out what you enjoy. Do further research into the tech roles you prefer, those that may interest you. Speak to people already working in those roles, as this is the best way to learn first-hand whether you would be likely to enjoy their day-to-day work or not. You can even ask to job shadow voluntarily for companies that are willing to allow interested people.

Do take risks and do not limit yourself to one area or line. Tech overlaps in many areas and is found in all different industries. When you're taking risks, do not be afraid to fail – this can be a great learning curve on your journey and only makes you better. Be focused, patient and keep learning.

FIVE

Tech Trajectory: Get Started, Now!

> 'We do not need magic to change the world, we carry all the power we need inside ourselves already: we have the power to imagine better.'
>
> —JK Rowling

It may seem like you need a magic wand to help you decide when you're thinking of what may interest you in such a vast industry as tech, but as JK Rowling so rightly said in her Harvard commencement speech, we already have the power to do everything we need inside ourselves. We can all dream and imagine, and then we can take a more pragmatic approach and plan our future. It's just about asking yourself the right questions, so I advise you to take some time and reflect a little during and after reading this chapter.

You have learned from the previous chapter about some of the most common job roles in tech, plus that it is possible to enter a career in this industry without really being technical. It is not always necessary to know coding. I would advise you to start your tech journey by thinking about what you enjoy as a hobby or what you were good at in school.

Take a minute and write a few things down. If nothing comes to you right away, do not worry. By selecting this book, you've proved you have an interest in tech, and I know you have it in you to take action and achieve your goals. Even if you are reading for interest only, you will learn skills here that everyone needs to succeed in any career.

If it is a specific role or area you may be interested in, then make a point to research more. Look at job adverts and job descriptions for the role you desire.

After you've identified your interests, I suggest you look at your past and current experience, including holiday jobs, work or academic experience, or even volunteer work. Review any feedback you received in the past as it can be a good indicator of your strengths. Please think about all your talents and skills; everything counts.

Once you are clear about which area you are most interested in, it is vital to focus on your career goals. Where do you want to put your ladder up against?

What is the reason you want to go into tech? What really excites you about the industry? What do you want to achieve in tech? Goals that truly excite you will ensure you keep going to achieve them.

Regarding job descriptions, it is good to get an idea of how big the gap is between your current skillsets and those required for the role you desire. For example, if the role is requesting intermediate Excel skills and you have never used it before, then you know the gap is large. Alternatively, if the job description requests proficient Excel skills and you are a basic user, it could be that you won't have too much difficulty learning some more.

I encourage you to apply for jobs regardless of whether you meet the criteria according to the job description or even if you have more to learn. However many boxes you tick for each skill required, please still apply. Studies show that females will not apply for a job unless they meet at least 78% of the criteria (*Harvard Business Review*, 2014). This has to change because if you never try to join the party, you will never get invited. Males are likely to apply even if only 60% of the criteria are met.

Training and learning routes

I mentioned earlier that people in tech benefit from constant learning, which as we saw in the previous section may need to start before you even enter the industry. Let's discuss formal education first.

As you know, I researched and looked at many options, including short courses, before I opted to study for a year. However, many hiring managers in companies are aware that a college or university degree is not a sign that a candidate will succeed in the role. Often, hiring managers complain that college graduates need more practical skills for the job and require a lot of training as newly hired recruits.

In my experience, the degree I chose gave me plenty of hands-on practical work through the assignments and teaching, so I gained the skills I would require. Some employers place more value on this practical work than class-type learning. This is something to think about when you are researching your entry route into tech.

It is also important to know the best way you learn new things. I preferred the route I chose purely because the course gave me all-round knowledge with practical experience so I could quickly achieve my goals. It was not about gaining the formal education of an MSc. In addition, it suited my circumstances at that time. The experienced tutors were helpful and always answered my questions with good examples from the industry, as quite a lot of them had worked in tech.

Also, because it was a conversion course, I had the opportunity to meet other students from all around the country – some with similar goals to mine and some with vastly different ones. This expanded my network and it was good to see all these different people bringing their experiences to the table. It was a surprise to me how many older students like myself were completing the course for different reasons.

Here are some useful questions to ask yourself when you are researching training options for a career in tech:

- Would you be able to commit to a programme or course for longer than a year?
- If a formal education route is not your preference, is it necessary for the area you wish to go into?
- Are you able to select a few courses to fill in any skills gaps?
- Do you prefer to take courses emphasising technical skills? Do you enjoy the practical side of things?
- Would you consider taking challenging courses that are not directly linked to your primary chosen technology area?
- Would debt be something you are willing to take on to pay for the cost of training? Could you rely on the better salary and prospects to pay off this debt when you secure your job at the other end?

I appreciate these days there are considerable costs involved in obtaining degrees, so consider your situation carefully. Look into sources of funding such as grants or schemes or voluntary sectors that may assist you.

Online learning

Many online courses are available these days on various platforms, such as Coursera (www.coursera.org) or Pluralsight (www.pluralsight.com). These give you the flexibility to study on the side while continuing in your current job or education.

Local colleges offer short courses and there are many women in tech networks who will guide you and offer lunch-and-learn sessions. You can refer to the resources at the back of this book for more.

Hackathon

This is a great way to build up skills in a short timeframe while meeting other like-minded people to widen your network. If you are new to programming, it is also a great way to learn from others who have more experience. There is always a good mix of people at hackathons.

A hackathon is an event to develop new approaches or new techniques for solving a problem, such as cybersecurity. It can be a few hours long, a day long, or even held over a few days. Those choosing a hackathon meet to see if they can break a program or hack into it, or maybe look at solutions to a problem in the programming. Hackathons can be coding challenges or business challenges, and there are often cash prizes for winning.

A hackathon is a beneficial way to learn more as you meet others with different experiences. You will often find that corporate companies hold these challenges, which is great for your essential networking – more about networking later. Eventbrite is an excellent place to search online for future hackathons in your area.

Boot camps

These are beneficial training camps to learn technical skills in a short period of time. For example, there are boot camps for learning coding, cybersecurity, AI and many more areas, and they are usually run by professionals working in the industry.

There are universities that run boot camps. There will be small fees involved, but it is a great way to learn new skills. However, be sure that the boot camp you choose covers what you are looking for, and attend the entire event to benefit fully.

Conferences and seminars

This is one of my favourites, and it is definitely a worthwhile investment as you learn so much and meet so many people in a short space of time. There is no shortage of conferences or seminars in tech, so there are bound to be plenty near you. Research these and always review the agenda before attending as there may be topics unsuitable for you, and there can be two or more workshops running at the same time, so ensure you select the relevant ones.

Regarding funding, you may be able to have your ticket sponsored, so your workplace or school pays the ticket fee. Often, large corporate companies sponsor conferences, so you have the benefit of meeting with representatives from these organisations and asking them questions. I have noticed many times when I am on a stand at a tech conference, representing the corporate company I work for, that people don't tend to take advantage of this opportunity. Aside from the freebies on offer, you do not know how much you can learn from people at these events so do stop and ask questions.

Volunteering

This is something that is often underestimated in relation to the benefits you gain. It is a great way to obtain experience, plus you help companies that may not have the resources to hire.

Of course, you must ensure the opportunity is aligned to the skills you are looking to acquire or improve. Also, check what kind of organisation it is you will be volunteering within. Is it a community-based organisation that may have events you can attend, or is it a charity?

You will want to be clear up-front on the time you're willing to offer and the length of your commitment, and ensure the company does not use too much of your time. I was lucky to find a volunteering opportunity in my local hospital's IT department, as I knew the staff there had a lot of work on and the department was expanding. This gave me invaluable experience and led to my first job. Never underestimate the power of doing some voluntary work because everyone benefits.

This section has given you some tips to help you get started, but you do not have to do all of them straight away. Choose a few you are comfortable with, focus on them, then move to another option so you ensure you take as many steps as you can to move you forward along your journey.

Personal branding

Looking back, when I started my career, I never thought I would be a multiple award winner or be invited to speak on careers panels and at tech conferences, or teach people and business owners new to tech, let alone give a talk on stage on International Women's Day. I have achieved all this because I have built up my brand, which means I am now approached for opportunities I enjoy outside the day job.

Today in the digital age, no matter which industry or profession you are in, building and maintaining your reputation – online or offline – is everything. Personal branding is often underestimated, misunderstood or perhaps not done as well as it could be, but if you want to be a techie woman, you can definitely give this a real kick and boss it. In this section, we cover areas of the corporate landscape that we should all be aware of to help you deal with building your brand more efficiently.

In her first book, *The Definitive Guide to Business Start-Up Success,* Bianca Miller-Cole says, 'Not enough people see the power in marketing themselves in the right way by being their best authentic self. How you communicate, your social media, how you dress, how you present yourself: all of that is part of your brand.' To start building your brand, think about *what* you want to be known for. What do you want to stand out for? What are your values? These are key to your brand.

Think deeply about the questions above. Because nothing is left in the dark when you are online, take care of whatever you share and post on social media. You always need consistent messaging on all platforms that you use. The *Harvard Business Review* tells us '70% of employers check out applicants' profiles [on social media] as part of their screening process, and 54% have rejected applicants because of what they found' (Wong, 2021).

A personal brand is a collective of your experience, your skills, your personality and what makes you 'you'. It is all about reputation and what people will see and say about you when you are not with them. Before you build your brand, you must be crystal clear about what you want to be known for. Another example is if you were to type your name into Google – what you would like to see about yourself.

For example, would you like to be referred to as a social media expert, an influencer in your field, a skilled programmer who troubleshoots problems and develops great websites, an excellent coder, a blogger, a public speaker, a thought leader in your specialist area of expertise, etc? Think about putting your name into a Google search. What information would you like to see come up about you? Whether you're getting hired for a job role or invited to interviews, people will look you up and see your online footprint before they have even met you.

When I used to prepare for interviews, when time allows, I always check LinkedIn to see if what is on the company's résumé matches and aligns with my own personal brand and values. This, among other reasons, is why you must control your narrative. Ensure you take care of your online reputation and have consistency in what you portray on all platforms. This is simply good practice. It is also good to check the other way around to match your values with the potential company you may be interested in.

I suggest you keep one of your social media profiles for personal content only, so you can share this with friends and family, and ensure your privacy controls are on the correct settings. For example, Snapchat, TikTok or Facebook are popular platforms for more personal content.

When you're invited for an interview, remember your appearance. Blogger Alysa Hinshaw tells us, 'Within seven seconds, the person across from you is assessing whether you're likeable, trustworthy, and competent' (Hinshaw, 2020). This makes things difficult as you often do not get a second chance.

In tech, many jobs have a casual dress code, so it is fine to wear jeans and t-shirt when you're with your team, but you can still make sure you look presentable when you need to convey the right impression, especially when you are new in the role. Find out before your initial interview what the etiquette is within a company, including the dress code, as you want to make a good impression from the start and you only have one chance to get this right.

When you're considering your brand, think longer term, beyond your ideal job or current job responsibilities. Where do you want to be in the future? How is the job or technology industry likely to grow over the coming years?

SheChoseTech has been my brand name since before I started writing this book, but you do not need to have a specific brand name. If you have a side-hustle business or are likely at some point to be self-employed, I recommend that you do, although your own name could be enough. I decided on SheChoseTech to create awareness and inspire people, especially women. You too need to be clear about what you want to be known for and make sure your brand reflects that.

Authenticity is crucial for your brand. Always be yourself and express yourself in the way you would normally, while making sure your message is clear and your followers will be left in no doubt what you mean. Often, I may read a post on LinkedIn and immediately realise the owner of the profile did not write it as I know them personally and recognise the way they speak. It is essential to remain authentic and let your personality come through.

LinkedIn

This is a great tool for people in work or in business. If you do not have an account already set up, I recommend you create one. With over 900 million users worldwide and 38 million in the UK, this is the largest professional networking site. It is your best resource to look for roles and network with the professionals you would like to speak and connect with.

Here, I would like to share some essential tips for using LinkedIn. For a start, you can control the narrative for yourself. On LinkedIn, you can find potential employers and senior decision makers, along with former colleagues or even people you may have studied with. There is a difference between your updated curriculum vitae and your LinkedIn profile, as you can also use LinkedIn for networking, connecting with people and sending messages. Managers, teachers, friends and colleagues, co-workers, mentors – there are so many influential people on this platform.

You can use it to post your work or write articles to showcase your professional expertise, so if you enjoy writing blogs or sharing information, LinkedIn is the perfect place to do so. You can also receive and display recommendations from other people for work you may have completed as part of work experience or in a job role.

Ensure you have a professional photo of yourself for your LinkedIn profile. You'll need a headshot only. While you may like spending time with your pets or your family, those pictures are better kept for the sites you socialise on. For LinkedIn, you need to be – and look – professional at all times.

LinkedIn offers a short voice feature where you can introduce yourself, if you feel brave enough. Anyone clicking on your profile can then hear as well as see what you do or which area you are interested in or seeking work in.

The headline on your profile, which is a short piece of information underneath your name that sums up your current professional status, is key. If you use this properly, it is one of the first things people see about you when you are searched, the crucial first impression. These key words used in the headline appear in searches. If you are new and using LinkedIn for the first time, I recommend you ensure your headline is updated regularly to reflect your role, or if you are studying, you could add 'Student completing XX course at YY University/College, seeking a full-time role in technology as a ZZ'. This gives a clear overview of you straight away, is searchable and lets prospective employers know what you are studying and where.

Make sure all the sections on your profile template are completed, as the more information you include, the better it will be for you when someone is looking at your profile. Connect with people by sending requests to current and former colleagues, teachers and friends. Make quality connections, rather than for the sake of it, as this is a professional platform. When you have completed all the sections, make sure you set your profile to 'public' so all can see you and search for you.

To expand your LinkedIn network, join groups of interest relevant to the tech industry. You will find many examples when you search for them. These groups are another great way to learn and network.

For example, if you are interested in project management, there are many different groups to search and join on LinkedIn. When you join, you will see plenty of posts from industry experts so you can continuously learn. Online networking on LinkedIn can be done through either messaging people of interest directly or liking and commenting on their posts to get noticed.

I have been approached many times in relation to roles by recruiters who found me on LinkedIn. In fact, that's how I entered my current job. I have also been contacted on LinkedIn to speak at events or to connect generally with other industry experts, which has resulted in productive conversations and quality connections.

Like all social media, LinkedIn is most beneficial when you post regularly, so do share professional updates and talk about what you are doing. If you read something interesting, share it with your followers and add your own comments and opinions. By posting regularly, you will encourage other users and your connections to start commenting on and liking your posts. Be generous with your posts, comments and knowledge, and you will grow your following.

As LinkedIn is a professional platform, think about who you wish to contact and make sure you benefit from quality connections. If you create a strong personal brand that clearly expresses who you are, what you stand for and how you are different, you are going to attract better opportunities. To increase the connection between ourselves as women and a future pipeline of leaders, we all need to be visible online and offline, and talk about our accomplishments.

Lastly, stay safe online. If any connection makes you feel uncomfortable, do not give away any personal information. In fact, I recommend you never share your personal information, as there are fake online profiles out there. On the whole, though, LinkedIn is an excellent tool to use. You can even use it in lieu of having your own personal website, which I did for a long time.

Networking

Over the years of my career, it has taken me a lot of time to build a great network; this is not created by accident. It is always good to connect, and networking in particular is about *give and take,* so it's extremely valuable as you progress through your career.

Don't ever view networking as a one-sided interaction for your benefit only. It's all about a win-win relationship, so never go to events or make calls just thinking about what the other person or people can do for you.

It's crucial to be clear on what you can offer them, too. Can you share something of use to them or introduce a contact that you know who specialises in something they may require help with? Can you share information about a new event you are attending or an interesting article you read in their area of expertise?

Many people believe the best jobs are found through strong ties they already have, like family and friends, but the truth is that the most valuable leads come through conversations with distant connections like friends of friends. Tap into a wide network as it is always a great way to learn and expand your connections. Plus, if you have a wide circle of connections, you learn new things from a diverse group of people, rather than the same likeminded colleagues that you regularly hang out with. Join a community to develop your network.

When you're networking, always follow up with the people you meet. Simply going to conferences or tech meet-ups is not sufficient, so have a plan on what you would like to do afterwards. There are many occasions where I have met people and we have exchanged business cards or connected on LinkedIn, and then I have never heard from them again, even though I have messaged them.

Ensure you follow up within a week of meeting people while it is still fresh in their mind who you are, and then build the rapport from there. For example, if you can share something with them, you may want to arrange a call or meet with them to discuss this and whether there's anything else you can offer that would be of value to them. Many people underestimate how much value there is in the follow-up.

Finally, keep in touch long term with the people you meet via networking, so when you genuinely need to reach out, they're not left with the impression you only connect when you require something. Build the rapport and relationship beforehand.

Networking is an important tool to have in your toolbox. Over time, you will increase your confidence, as I appreciate it is not easy to speak to strangers. It takes time and practice. Always have your opening lines ready and rehearsed so you can articulate what you do. It is rather like having a one-minute elevator pitch ready to meet the world.

If you find this preparation or anything else about networking difficult, do seek a mentor, someone with experience that you admire. They can guide you and share helpful tips on what to say when you message or speak with people face to face to open meaningful conversations. Do not forget to listen really carefully to the people you meet so you can remember them and converse effectively.

Imposter syndrome

You may have heard of the term *imposter syndrome*. This is the antithesis and destroyer of confidence, so I would like to share this important point here as it affects many successful people, both women and men. As you move through your career, or in anything you may do, imposter syndrome is likely to come and go, depending upon the situations you find yourself in.

I have experienced imposter syndrome from time to time. For example, years ago when I was requested to chair an important meeting for my manager, I immediately questioned my ability – without knowing the details. How could my manager think I would be able to cover for them?

Had I truly believed that I was enough and knew enough, then the awful looming dread of imposter syndrome would not have crept in. However, if I had lacked the skills – if the manager had not thought that I was capable of chairing the meeting – the request would not have been made of me.

Hira Ali in her book *Her Way to the Top* shares articulately what imposter syndrome is: 'We all feel inadequate at some point in our lives. Often it's when we don't recognize our own self-worth or believe we are qualified enough to achieve something. It's natural to feel "not good enough" when we are pushed outside our comfort zone or trying something for the first time.'

The term *imposter syndrome* was first coined in 1978 by clinical psychologists Dr Pauline Clance and Dr Suzanne A Imes to describe a persistent fear of being marked as a 'fraud', and has remained in use ever since. Reshma Saujani, an American lawyer who advocates for women in tech, leadership roles and STEM, and is the founder of the non-profit Girls Who Code which has educated more than 300,000 girls in computer science classes since its founding in 2012, is no stranger to this insidious self-doubt. She gave a prominent speech to graduates at Smith College in the United States about how she learned to deal with imposter syndrome, which is something many females contend with. In it, she said, 'Imposter syndrome is based on the premise that we are the problem, that if we feel underqualified, it's because we are. If we worry that we don't have what it takes, it's because we don't' (Saujani, 2023).

If you read about her, you will find that Reshma Saujani had experienced feelings of inadequacy and often felt alienated growing up as the daughter of refugees in the United States. In spite of this, she has become extremely successful, yet still she is plagued with self-doubt.

There are many who keep stating that high percentages of women suffer from imposter syndrome. At times, women will encounter the feeling that they don't fit in, and this, to an extent, is normal. However, Saujani suggests that this will never change if it remains a focus point of constant discussion – she believes imposter syndrome may be a distraction. If we as women are constantly worrying that we may be inadequate or frauds in some way, even when the evidence is that we are anything but, it keeps us from focusing on sexism, racism, classicism or any other prejudice we may face. Reshma Saujani concludes, therefore, that it's the system that needs fixing, not ourselves. Our focus needs to be on healing what appears to be a broken world. If we are to move the dial at all, it's helpful to keep this front of mind. Over time, Reshma Saujani says, she realised her imposter syndrome was not a reflection of her own abilities, but a much larger problem: inequality.

Readers, do take note of this. I have lost count of the number of times I have seen sessions organised with speakers to answer questions on dealing with imposter syndrome. This is a good time to reflect on Reshma Saujani's words and start changing things. She has a valid point here, because so many of us with imposter syndrome believe something needs fixing in us.

Another related point I will make here is that we often talk about male allies. These are the men among our colleagues who help with the problems we women may face in the workplace. A male ally stands in solidarity with women, actively working to dismantle gender inequality.

While I completely advocate for male allies, I want to ask women to work together more. Have you considered helping yourself and other women first? There are so many examples where I have seen and experienced women not coming together and supporting each other enough. This is a point I share many a time through my role mentoring or coaching women. I have repeatedly heard stories about workplace problems with other women, which could so easily be avoided if only we females all worked together and not against each other, either in competition or otherwise.

It is never good practice to gain what you can for yourself while not thinking about your colleagues. You may have noticed how men often seem to hit it off with each other over conversations about sport or current affairs, or simply chatting over a pint. Do we as women need to do something similar, to stop competing with each other and help our female colleagues rise? We expect male allies to be there for us, but first and foremost, are we doing enough for ourselves? Keep this thought with you as you embark on your tech career.

I have worked in employee resource groups or networks in the corporate world. This is where I come together with others to organise exciting internal events or sessions with talks on careers or topical issues, maybe interviews with senior leaders for all to benefit from – all on top of the day job. This is a clear example of how teamwork and cooperation can help everyone rather than women having their own agendas to elevate themselves while treading on other females in a distasteful way. Always think about how you can help in your networks and give back to communities. We women would be much further ahead if we stood together more consistently with almighty 'girl power'.

Take action today and find yourself an accountability partner. This can be a trusted colleague or friend. List all the examples that come to mind where you can improve the world of tech as a woman. In the meantime, I'll leave this section on this quote from Reshma Saujani about imposter syndrome: 'It's not your job to fix yourself. But it is your job to fix the system.' (Solow, 2023)

Self-promotion or how to toot your own horn

Self-promotion will come with experience and practice, as it is something that many people find difficult to do. It goes back to developing your brand and being clear on what you would like to be known for, so think of three words to describe yourself as a start. What makes you stand out, and when you are networking, what do you want to be remembered for?

To help promote yourself or toot your own horn in the nicest possible way, I recommend you keep a list of all your achievements and successes. It can be as simple as putting a few quick notes into your phone and using this list to remind yourself of everything you have achieved, for work as well as outside your career. On down days, and we all have them occasionally, you can refer to it and recap on all the wonderful things, big or small, that you are proud of.

Ask a friend to share something positive about you if you cannot recall anything yourself. State facts as they are, as we can get too emotional when things mean a lot to us. Always give praise and credit to people who have helped you where it is due. This is important because as you spread the good vibe, it is returned.

Accept praise with a simple thank you. So many times when I praise someone, they almost dismiss their achievement and rub it all away with, 'Oh, that old thing.' Thank people for every compliment they give, do not wave away the credit. This applies even more when you are at work.

When you are showcasing an achievement, always put in a little story about your struggles or difficulties. Be authentic, as people love to hear your stories and relate to your vulnerability. If you are not authentic, it is clearly seen by all.

You will find that the skills I have described above capture a lot of the main areas of self-promotion experienced by people throughout their career, so it is important to start to hone these skills early. Seek support through your colleagues and friends to progress, as it takes time. For example, to have in place personal branding that you are truly comfortable with so you can use LinkedIn proficiently, ask a friend or colleague for advice on what works. Your career is a journey, and it is fine to make mistakes and learn from them.

It is important to find a mentor to support you on your journey. This is a person with more experience than you and someone you may admire for a skill they have. You ask them questions informally to help you learn; the arrangement does not need to be formal. You simply meet and discuss your questions and issues regularly with your mentor. It is also important to be aware of what a sponsor can do for you, which is someone who advocates for you when not in the room.

This chapter has covered some key ways into a tech career and areas to be aware of when building your personal branding, such as using LinkedIn and networking, and being authentic. All of these are valuable areas to learn or improve upon as they will stay with you and be of use throughout anything you choose to do.

It is important to be aware of imposter syndrome. The valuable opinions I have shared in this chapter give you a starting point to take action against this negative syndrome that afflicts so many of us. Be especially mindful to support other women and research about allyship as this is another crucial area to consider on your journey.

It takes practice to remind yourself of all the positive achievements, big or small, that you have in your life. None of us self-promotes enough. To increase your confidence, keep a list in the back of your notebook or on your phone of all the wonderful things you have done and continue to do.

In the final chapter, I will share some important points on general wellbeing. There is often a work hard, play hard culture these days in any career. An important subject we'll cover together with wellbeing is DE&I. In the next chapter, you will hear more about this and what you can do to ensure true DE&I in your workplace.

SIX

Tech Team: Fostering DE&I And Wellbeing At Work Through The Power of Diversity

> 'Our ability to reach unity in diversity will be the beauty and the test of our civilisation.'
> —Mahatma Gandhi

Mahatma Gandhi fought tirelessly for the rights of everyone in his society. As such, he was a man who epitomised what it means to respect and prioritise diversity in all walks of life. His words express beautifully how important DE&I are and will continue to be into the future.

My final chapter applies to everyone in today's workplace. Corporate bosses increasingly recognise the significance of staff wellbeing, and they understand a healthy and *engaged* workforce is productive. For me, the true potential of a thriving work environment is unlocked when wellbeing is coupled with diversity, equity and inclusion (DE&I).

This means *everyone* being included, regardless of their ethnic background, religion, sexuality, gender, etc. It should not take another Black Lives Matter campaign for the focus on DE&I to become essential. This needs to be a part of everyday life so everyone feels comfortable to be themselves and bring their best selves to work.

It's obvious when an organisation prioritises employees' physical, mental and emotional wellbeing and holds this in high regard. These employees are more likely to perform at their best than those whose wellbeing is not a priority, leading to increased productivity and, ultimately, profitability for the company and everyone within it.

The modern workplace has witnessed a shift in societal values, especially since the Covid pandemic, so there is a greater emphasis on work-life balance, mental-health awareness and employee satisfaction. This shift in values points to senior leaders and all managers who understand that attracting and retaining top talent, importantly diverse talent, requires creating an environment that supports and nurtures the wellbeing of their staff. Ultimately, by prioritising staff wellbeing, senior leaders are not only investing in the success of their organisation, they're also acknowledging the worth and dignity of the individuals who contribute to its growth.

I have made it clear throughout this book that we need to increase the ratio of women to men in the tech industry and ensure those women who are recruited are drawn from as diverse a range as possible, because diversity of thought matters in all tech developments. In fact, this is true in all organisations. Imagine what could be if we were consistently cultivating wellbeing at work and focusing on embracing DE&I.

What is DE&I?

Diversity encompasses differences in gender, race, ethnicity, age, religion, visible and non-visible disabilities, and more. To embrace diversity means we recognise and value these differences and ensure people from different backgrounds are represented and included. As Sheree Atcheson says in her book *Demanding More*, 'Diversity in groups allows the blend of perspectives, experiences and ideas, on the same topic or conversation. When we live, socialise or work with people from all different backgrounds, we ultimately hear perspectives that we may never have heard of before.'

Equity is now being talked about more than it used to be, and so it should be. Equity seeks to create a fair society by addressing historical disadvantages and barriers people may have faced. It is about ensuring all have the same access to opportunities, resources and treatment.

For example, this means all employees have fair and equitable access to training and development to further their careers. Or this could mean students from disadvantaged communities receiving extra help or resources to bridge the gap in achievement, so that they all have the same educational opportunities as students from more fortunate backgrounds. However, as we have discussed in this book, there are many different routes into the tech industry, so you do not always need to have an undergraduate degree. For those who want to study for a degree, the opportunity should be open to them, and that is one thing that the principle of equity addresses.

Inclusion is where all individuals feel welcome, respected and valued, and their contributions are acknowledged. This can enhance employee satisfaction, foster innovation and create a more inclusive and harmonious workplace culture.

By promoting DE&I, organisation leaders validate their employees' unique identities and experiences. This recognition creates a sense of belonging, which is vital for overall wellbeing. When individuals feel seen and accepted for who they are, they are more likely to thrive, be engaged and perform at their best. DE&I allow a blend of different perspectives and experiences and help to get the best from a wide range of understanding and shared ideas. This rich tapestry of different perspectives expands collective knowledge, encourages critical thinking and fosters creative problem solving.

Ruchika Tulshyan, an award-winning inclusion strategist and best-selling author who speaks a lot about everyday biases, has worked in tech and did not see diverse, inclusive or equitable workplaces around her. She says, 'As leaders, we need to become aware, intentional and practice inclusion every day, in every moment. We have to overcome individual defensiveness to focus on learning and growth' (Taylor, 2023). This practice needs to be adopted by everyone. Learning where our biases are is a good place to start. We need to try to be mindful of our own biases (and we all have them, consciously or unconsciously).

In your day to day, if you are comfortable with it, it is good to challenge the status quo. This is something everyone should learn to do in the best possible way. Challenging yourself not to have too much ingrained thinking is important to see equal treatment for all so you can level the playing field. This is something you can encourage in your networks while ensuring you all work together and support one another.

When my company is hiring and I am asked to interview candidates, I ensure there is a diverse interview panel. This helps interviewers to challenge their own biases, where they can be challenged, during the hiring process. All people gravitate towards others similar to themselves – especially when hiring.

It is also beneficial to look at job descriptions and work with HR to check they use inclusive language that encourages a diverse mix of women to apply. This is not because we wish to see special treatment for women in tech; of course, what we would like to see is equality.

Mentors

In parallel with promoting DE&I, in our day to day, it is beneficial for us all to become more aware of how to look after our own and our colleagues' wellbeing. To help with this, I suggest you find a mentor who can use their experience to guide and support you. Mentors don't only help for your career goals, they also enable you to learn more about diversity and different walks of life.

I appreciate that communicating with a diverse range of colleagues can be awkward, as many of us don't always know what the right or correct thing is to say. For example, do you say women of colour or black women? It is good to engage with someone who can guide you and be specific when you talk about these issues as sometimes, as in the example above, neither is wrong.

When you're seeking a mentor, be clear about what you want from them before you engage with someone, and have goals in mind for what you would like to achieve. You may find many formal mentoring opportunities in your networks outside of the workplace that you can take part in.

The women you have heard from in this book have had mentors, and so did I. It is always useful to have someone to advise you over a problem or answer your queries on how to move through the corporate landscape, including changing roles, promotion, DE&I, wellbeing or anything else you get stuck on. You then learn from someone more experienced than you. Of course, when you do come to the end of your mentoring journey, it is important to thank your mentor for all their help and guidance.

In addition, to make your thanks more powerful, I suggest you pay it back by becoming a mentor yourself, once you are comfortable doing so. This way, you can help guide other people just as you were guided, as by this point, you will have trodden their path. By sharing our experiences in this book with you, including the interviews, I have shown how easy it is to pay it back when you have benefitted from mentoring.

Mentoring is important. It means each of us can influence change little by little. As TV presenter June Sarpong says in her book *Diversify*, 'Even in the smallest of ways, we can all do our bit to boost diversity and promote a peaceful environment at work.'

Wellbeing

When things get difficult, taking care of people's wellbeing is important. This is not just about physical health; it encompasses mental and emotional wellness, too. By embracing DE&I, we bring together individuals with different backgrounds, beliefs and experiences in an environment where they all feel safe and comfortable to be themselves.

During Covid I took part in a panel discussion sharing how I dealt with my personal loss and grief which occurred just before Covid. I believed it was important to share this with the audience. I was comfortable to say yes when asked to speak on the panel, as I thought it was important for others to be aware about different challenges we all go through and come out from. This only occurred because the work environment allowed for it. It is also important to speak about mental wellbeing and different coping mechanisms.

I find that when employees engage in meaningful discussions and learn from each other's diverse viewpoints, they experience personal growth and fulfilment, enhancing their overall wellbeing. I have had this experience through workshops, panel events and talks I have taken part in or organised. It is so good to see the many who benefit from these kinds of interactions.

When employees feel safe to express their authentic selves without fear of any discrimination, subtle or obvious, it promotes psychological safety, trust and emotional wellbeing. If you encourage open dialogue and address biases within your organisation, it fosters a culture of respect and empathy, benefiting everyone involved.

In addition, those who feel valued and supported are more likely to be satisfied with their work – leading to higher levels of engagement and productivity. When DE&I and wellbeing are prioritised, employees tend to remain loyal and committed to their jobs. This is important for the retention of talent, including female talent, in the tech industry. There are many organisations that place importance on wellbeing, so if yours does not, do not wait. Suggest ideas for improvement to line managers (you'll find plenty of ideas in this chapter) and help make your environment better for all.

Innovation is like the lifeblood of any successful organisation, and diverse teams are more innovative as they benefit from such a wide range of inputs. A diverse workforce brings together individuals with unique experiences, skills and perspectives, which fuels creativity and leads to novel ideas. When employees feel comfortable sharing their ideas without judgement, this cultivates a culture of innovation that not only benefits the organisation, but also nurtures the individual wellbeing of employees who thrive in such an environment.

If employees are encouraged to bring their whole selves to work and are supported in their uniqueness, they experience a sense of belonging and fulfilment. Through fostering an inclusive culture, organisations can drive positive change, challenge biases and create a workplace where everyone can thrive. The path to a prosperous future lies in recognising the immense importance of wellbeing at work with diversity as its cornerstone.

INTERVIEW WITH A PROFESSIONAL – AKUA OPONG

Akua Opong is a keen DE&I advocate, a mental-health champion, community action lead for the UK Charity Committee, LSEG's UK accessibility lead, and community lead for both Women's Inspired Network (WIN) and WIN Tech. She is passionate about raising the profile of women in STEM as a STEM ambassador. Akua is a neurodiversity advocate (she self-identifies as dyslexic and has attention deficit/hyperactivity disorder (ADHD)), an avid fundraiser, plus a charity volunteer and a keen sports volunteer for Netball England and British Athletics.

She says, 'As someone who grew up in a low-income community, I am acutely aware of the lack of representation of women, especially black women, in the tech industry. I have found myself as the only woman or black person in too many rooms, which has fuelled my passion for empowering other women and driving change.

'At points in my career journey, I have had to work part time alongside full-time roles to make ends meet. These experiences have taught me to be resilient, driven and committed to my goals. As I continue to pursue my career in the technology industry, I am enthusiastic about the prospect of making a direct contribution to organisations that champion DE&I.

'I firmly believe that by increasing the representation of women, especially black women, in leadership positions, be it in the workplace or as non-executive directors (NEDs), we can create more equitable and inclusive workplaces that benefit all employees and the communities they serve. I am deeply committed to driving change by inspiring and encouraging other underrepresented women to actively pursue leadership and board roles.

'My aspiration is to use my experience and capabilities to make a meaningful impact on the technology industry at all levels. I hope to serve in senior leadership roles and boards in the future, while also inspiring and empowering other women to pursue technology roles. Women deserve a seat at the table and should be part of the important discussions in the workplace and valued more.'

Akua highlights some important points here. You will find a lot of inspiring women in the tech industry, but we need more. If we all, as leaders, continue to challenge and make changes in this industry, we will leave a better future and legacy for all. She also demonstrates many different sides of DE&I in her response.

While tech can be creative, artful, inspiring, dynamic and fun, a crucial piece of advice is to look after your wellbeing. Tech also brings challenges and often pressure to be constantly online or connected.

Between studying or work, there is a requirement to look after ourselves to ensure good balance and avoid blurring the lines between our day-to-day and home life. This is so that we may perform at our optimum, whether we're at school, college, work or business, to achieve personal and professional goals and, of course, to be there for friends and family. It's important to feel and be our best. It is *not* fine to be online or connected all the time, so give yourself a break.

When you ensure that you maintain a balanced life by separating non-work and work commitments and activities, it helps with creativity as the brain is given space. Creativity is essential in the tech industry to remain ahead with new solutions and innovations.

In this chapter, we have explored how employees' wellbeing, coupled with DE&I, is so important. Both need to be considered together for the best outcomes for all. When a diverse range of employees is working well together, this creates a safe environment for everyone to give their best, and the outputs are high productivity, innovation and job satisfaction. Everyone benefits in the organisation, which aids retention.

We also covered how important it can be to get a mentor to help you on your journey. Even more important is paying the mentoring you received back once you're further along the path to help others benefit from your experiences. In this way, you ensure that the future of the tech industry will be healthy, diverse, inclusive, equitable and innovative.

Conclusion

This book has shown that the tech industry has a considerable range of opportunities available for new people to join, no matter where they are in their life stage. The demand for talent in tech and the relatively low percentage of women compared to men indicate that once this is addressed, it will benefit the industry. It will also help create considerably more tech leaders who are representative of the diversity of people the industry serves.

Global demand for innovation in tech is growing rapidly with advances in AI, cryptocurrencies, blockchain and the need for sustainability. These are all driving the industry to close the gender and diversity gaps in recruitment. Diversity of thought, which is essential in the developmental processes of new technology, needs women's voices.

To bring you into the tech world if you're new to the industry, Chapter 3 describes a few of the many emerging, current and popular technologies, such as AI, cryptocurrencies and blockchain, to help develop your knowledge. AI in particular needs to be used wisely to release its potential to support human development and advance benefits for humanity.

It is crucial for everyone to be mindful of how computers work and how the biases in the sector can affect AI, another reason why we need more awareness and diversity in the profession. Chapter 3 also addresses a prevalent fear that humans will soon be losing jobs to AI, but these more traditional forms of employment will be replaced with exciting new roles, achieving a net positive.

In Chapter 4, I shared some roles that you will typically find in tech, with details about what each role involves day to day. You heard from some women who are already rocking these roles, describing why and how they joined tech and sharing their thoughts and opinions on the industry. We also examined qualifications, training and different routes into tech, as one size does *not* fit all. There are many routes by which women can enter tech.

In Chapter 5, we looked at finding your strengths. Everyone can take something of benefit away from this topic. Chapter 5 also covered the joys of networking, personal branding and LinkedIn, plus tips on how to toot your own horn. We looked at a different way of being aware of imposter syndrome and what we ideally should be doing about this.

This range of topics will help with your career journey, no matter where you start or what you decide. Chapter 5 emphasises, from the examples and real-life experiences it shares, that anyone exploring a new career can be confident that joining the tech sector is possible for them.

The closing chapter covers the subject of DE&I, along with wellbeing, and discusses how important both are in today's working world. We should all work together to draw out our synergies.

If you are unsure where to start, I cannot recommend highly enough that you find an experienced mentor to guide you. Then when you're further down the road, you can pay back that help and guide someone else who is just starting out. Please also give back to others by sharing this book widely when you finish it and have absorbed the learning from it into your day to day.

For anyone, especially women, finding that dream job in tech will make you feel more self-assured, and you can then demystify tech to help others feel confident, too. This will shape and reinforce the benefits of a diverse workforce without leaving anyone behind, as the future has room for us all to grow the pipeline. Ensure you stay focused and empowered, and show that inclusive leadership will help us all to achieve considerably more, together.

You have heard from a number of women in the book about their experiences in tech. One reason women are underrepresented in technology may be due to their school years. Perhaps as girls, they weren't encouraged to study STEM subjects. Perhaps because the tech industry is so large, detailed careers advice is difficult to obtain for it, so women and girls simply don't know that skills other than STEM are also crucial. I hope the content of this book and the resources I have shared will go some way to addressing this if it is an issue for you.

We have much to do to help solve the lack of DE&I within tech. It is essential that we start to act now, because diversity of thought matters to all – to companies and the people who make them work, whether they are in tech or not.

If you are a woman, particularly a woman of colour, your voice is influential and needs to be heard. Your contribution is essential. We stand on the shoulders of giants, and tomorrow, many of us could be seen as the giants who helped others to contribute to the tech sector and make a real difference.

Do send me your feedback and thoughts on the book or share stores and ask any questions. Or for further advice use the contact details at the end of the book. Keep in touch and I look forward to hearing from you.

References

'About LinkedIn', LinkedIn (no date), https://about.linkedin.com, accessed 19 September 2023

Ali, H, *Her Way to the Top: A guide to smashing the glass ceiling* (Panama Press Ltd, 2019)

Ali, H, *Her Allies: A practical toolkit to help men lead through advocacy* (Neem Tree Press Limited, 2021)

Atcheson, S, *Demanding More: Why diversity and inclusion don't happen and what you can do about it* (Kogan Page Limited, 2021)

Bartlett, S, 'E252, Emergency episode The Diary of a CEO – Ex-Google Officer Finally Speaks Out On The

Dangers Of AI', YouTube (1 June 2023), www.youtube.com/watch?v=bk-nQ7HF6k4, accessed 28 August 2023

Bell, E, 'A fake news frenzy: Why Chat GPT could be disastrous for truth in Journalism', *The Guardian* (3 March 2023), www.theguardian.com/commentisfree/2023/mar/03/fake-news-chatgpt-truth-journalism-disinformation, accessed 23 August 2023

Chen, Nikol, 'Nothing is to be feared, only to be understood', Laidlaw Scholars (20 April 2020), https://laidlawscholars.network/posts/nothing-is-to-be-feared-only-to-be-understood, accessed 10 August 2023

Criado Perez, C, *Invisible Women Exposing Data Bias in a World Designed for Men* (Vintage, 2020)

'CSW67', UN Women (2023), www.unwomen.org/en/csw/csw67-2023, accessed 5 March 2023

Curie, M, 'Atomic Energy Review, Vol. 6 (1968), No. 1' [Centenary Lectures], International Atomic Energy Agency (1968), www.iaea.org/publications/1933/atomic-energy-review-vol-6-1968-no-1, accessed 28 August 2023

Davis, P; Carney, S, 'Katherine Johnson (1918–2020): Former NASA Research Mathematician', NASA Solar System Exploration (18 July 2023), https://solarsystem.nasa.gov/people/434/katherine-johnson-1918-2020, accessed 19 July 2023.

'Diversity in Tech 2021 report: An annual report tracking diversity in technology across the UK', Tech Talent Charter (2021), https://report.techtalentcharter.co.uk/diversity-in-tech-2021, accessed 1 December 2021

'Diversity in Tech: An annual report tracking diversity in technology across the UK', Tech Talent Charter (2023), https://report.techtalentcharter.co.uk/diversity-in-tech, accessed 4 May 2023

Dweck, CS, *Mindset: How you can fulfil your potential* (Ballantine Books Trade, 2006)

'Executive Summary', Tech Nation (no date), https://technation.io/diversity-and-inclusion-in-uk-tech/#executive-summary, accessed 20 December 2021

'Future of Jobs Report 2023', World Economic Forum (May 2023), www3.weforum.org/docs/WEF_Future_of_Jobs_2023.pdf, accessed 30 May 2023

Gandhi, M, *Peace: The Words and Inspiration of Mahatma Gandhi* (Blue Mountain Arts, May 2007)

Hinshaw, A, 'Seven Seconds to Make a First Impression – Make it Count!', The Center for Sales Strategy (5 October 2020), https://blog.thecenterforsalesstrategy.com/seven-seconds-to-make-a-first-impression, accessed 28 August 2023

Hymas, C, 'Home Office under fire over face recognition technology that fails to recognise very dark or light faces', *The Telegraph* (10 October 2019), www.telegraph.co.uk/politics/2019/10/10/home-office-fire-face-recognition-technology-fails-recognise, accessed 28 August 2023

Imafidon, A-M, *She's In CTRL: How women can take back tech – to communicate, investigate, problem-solve, broker deals and protect themselves in a digital world* (Penguin Random House UK, 2022)

'Is it time for a UK Accountability for Algorithms Act?', Institute for the Future of Work (10 June 2022), www.ifow.org/news-articles/time-uk-algorithmic-accountability-act, accessed 30 December 2022

Khan, A, 'Girls Who Code Founder Reshma Saujani Breaks Down the Biggest Myth About Imposter Syndrome', Inc. (24 May 2023), www.inc.com/alyssa-khan/girls-who-code-founder-reshma-saujani-breaks-down-the-biggest-myth-about-imposter-syndrome.html, accessed 19 September 2023

King, C, 'Take Five: At the current rate of progress, no equal pay until 2069', UN Women (24 February 2017), www.unwomen.org/en/news/stories/2017/2/take-five-chidi-king-equal-pay, accessed 23 August 2023

Krauss, R, '15 unsung women in tech you should know about', Mashable (8 March 2018), https://mashable.com/article/unsung-women-in-tech, accessed 2 January 2023

Lee, JAN, 'Unforgettable Grace Hopper', *Reader's Digest* (October 1994)

'Number of LinkedIn users in the United Kingdom (UK) from March 2020 to June 2023', Statista (21 July 2023), www.statista.com/statistics/1314310/uk-linkedin-users, accessed 1 August 2023

Marr, B, 'The Problem With Biased AIs (and How To Make AI Better)', *Forbes* (30 September 2022), www.forbes.com/sites/bernardmarr/2022/09/30/the-problem-with-biased-ais-and-how-to-make-ai-better, accessed 9 September 2023

Miller-Cole, B, *The Definitive Guide to Business Start-Up Success* (John Murray Learning, 2017)

Mohr, TS, 'Why Women Don't Apply for Jobs Unless They're 100% Qualified', *Harvard Business Review* (25 August 2014), https://hbr.org/2014/08/why-women-dont-apply-for-jobs-unless-theyre-100-qualified, accessed 10 October 2022.

'Innovation and prevention of violence against women', UN Women (2023), www.unwomen.org/en/digital-library/publications/2023/07/innovation-and-prevention-of-violence-against-women, accessed 19 September 2023

'Internet users, UK: 2020', Office for National Statistics (6 April 2021), www.ons.gov.uk/businessindustryandtrade/itandinternetindustry/bulletins/internetusers/2020, accessed 1 May 2023

'People and Skills Report 2022', Tech Nation (2022), https://technation.io/people-and-skills-report-2022, accessed 21 June 2023

Philp, C (MP), 'New digital strategy to make UK a global tech superpower', The Department for Culture, Media, Digital and Sport, Gov.uk (13 June 2022), www.gov.uk/government/news/new-digital-strategy-to-make-uk-a-global-tech-superpower, accessed 1 June 2023

'Public spaces need to be safe and inclusive for all. Now', UN Women (2023), https://www.unwomenuk.org/safe-spaces-now, accessed 24 August 2023

Rowling, JK, 'Text of JK Rowling's speech', *The Harvard Gazette* (5 June 2008), https://news.harvard.edu/gazette/story/2008/06/text-of-j-k-rowling-speech, accessed 26 August 2023

Saran, C, 'House of Lords launches an investigation into generative AI', ComputerWeekly.com (7 July 2023), www.computerweekly.com/news/366544137/House-of-Lords-launches-an-investigation-into-generative-AI, accessed 11 July 2023

Sarpong, J, *Diversify: Six degrees of integration, because the world is separate enough* (Harper Collins, 2017)

Saujani, R, 'Imposter syndrome Is A Scheme: Reshma Saujani's Smith College Commencement Address', YouTube (20 June 2023), www.youtube.com/watch?v=BoHDDgeQtlc, accessed 26 August 2023

Schieber, P, 'The Wit and Wisdom of Grace Hopper', OCLC Newsletter 167 (March/April 1987), www.cs.yale.edu/homes/tap/Files/hopper-wit.html, accessed 24 August 2024

Shepherd, J, '41 Essential LinkedIn Statistics You Need to Know in 2023', Social Shepherd (15 May 2023), https://thesocialshepherd.com/blog/linkedin-statistics, accessed 16 May 2023

Solow, B, 'Graduates are "Uniquely Qualified" to Make the World Better', Smith College (21 May 2023), www.smith.edu/news/commencement-ceremony-2023, accessed 26 August 2023

Taylor, V, 'Ruchika Tulshyan outlines how to practice inclusion with intention, and it's about to get reflective', LinkedIn (20 March 2022), www.linkedin.com/pulse/ruchika-tulshyan-outlines-how-practice-inclusion-intention-taylor?trk=articles_directory, accessed 1 June 2023

Tupper, H; Ellis, S, *The Squiggly Career* (Penguin Books, 2020)

'Women in Tech Survey 2023', Women in Tech (2023), www.womenintech.co.uk/women-in-tech-survey-2023, accessed 12 January 2023

Wong, R, 'Stop screening job candidates' social media', *Harvard Business Review* (September 2021), https://hbr.org/2021/09/stop-screening-job-candidates-social-media, accessed 28 August 2023

Wyman, A, 'The women who changed the tech world', Global App Testing (no date), www.globalapptesting.com/blog/the-women-who-changed-the-tech-world, accessed 24 July 2023

Resources

Angelini Pharma, 'Towards inclusion: Manifesto for diversity & inclusion' (no date), www.angelini-pharma.com/our-responsibility/diversity-inclusion/our-commitment, accessed 18 September 2023

Arnold, G, et al, *Women in Tech: A practical guide to increasing gender diversity and inclusion* (BCS Learning and Development, 2021)

Arruda, W, 'The most damaging myth about branding', *Forbes* (6 September 2016), www.forbes.com/sites/williamarruda/2016/09/06/the-most-damaging-myth-about-branding, accessed 18 September 2023

Bennett, N, *Managing Successful Projects with Prince2* (Stationery Office, 2017)

Bentley, C, *Prince2: A practical handbook* (Butterworth-Heinemann, 2002)

Chang, E, *Brotopia: Breaking up the boys club of Silicon Valley* (Portfolio, 2018)

Cole, B, 'Women Empowering Tech: Inspiring Stories of Success and Impact', *Forbes* (13 September 2023), www.forbes.com/sites/byroncole/2023/09/13/women-empowering-tech-inspiring-stories-of-success-and-impact, accessed 22 September 2023

Department of Economic and Social Affairs, 'Sustainable development: The 17 goals', UN (no date), https://sdgs.un.org/goals, accessed 18 September 2023

'Diversity and Inclusion in UK Tech', Tech Nation (2021), https://technation.io/diversity-and-inclusion-in-uk-tech/#executive-summary, accessed 18 September 2023

Frankland, J, *In Security: Why a failure to attract and retain women in cybersecurity is making us all less safe* (Rethink Press, 2017)

Free Tech Books: https://freetechbooks.com, Institute for the Future of Work: www.ifow.org

Imafidon, A-M, 'Women Tech Charge', *The Evening Standard,* https://podcasts.apple.com/gb/podcast/women-tech-charge/id1454224152, accessed 18 September 2023

Marschhausen, J, 'Life in focus', blog post (2018), https://lifeinfocus.me/2018/12/28/we-do-not-need-magic-to-change-the-world-we-carry-all-the-power-we-need-inside-ourselves-already-we-have-the-power-to-imagine-better, accessed 23 July 2023

Mayor of London Assembly, 'Mayor launches initiative to improve diversity in the tech sector' (29 November 2021), www.london.gov.uk/press-releases/mayoral/mayoral-initiative-to-improve-diversity-in-tech, accessed 8 December 2021

McQueen, L, 'S2 EP4 with Sonal Shah', The Talent Waste Show, https://podcasts.apple.com/gb/podcast/s2-ep4-with-sonalshah/id1561977967?i=1000597543363, accessed 18 September 2023

'People and Skills Report', Tech Nation (2022), https://technation.io/people-and-skills-report-2022, accessed 18 September 2023

Robinson, J, 'Ex-Google tech guru who issued chilling AI warning hired by Dragons' Den star Steven Bartlett', Business Live (1 June 2023), www.business-live.co.uk/technology/ex-google-tech-guru-who-27027941, accessed 10 July 2023

'Salary Survey Review 2023', Robert Walters Group (2023), www.robertwaltersgroup.com/news/expert-insight/salary-survey.html, accessed 9 January 2023

Stemettes Zine, https://stemettes.org/zine, accessed 18 September 2023

Thomas, P, 'The most damaging phrase in the language is "We've always done it this way"', LinkedIn (2017), www.linkedin.com/pulse/most-damaging-phrase-language-weve-always-done-way-peter-thomas, accessed 18 September 2023

Tockey, D, and Ignatova, M, 'LinkedIn Gender Insights Report' (no date), https://business.linkedin.com/content/dam/me/business/en-us/talent-solutions-lodestone/body/pdf/Gender-Insights-Report.pdf, accessed 10 December 2022

UN Women, 'Equal pay for work of equal value' (no date), www.unwomen.org/en/news/in-focus/csw61/equal-pay, accessed 18 September 2023

UNESCO, 'Guidance for generative AI in education and research' (7 September 2023), www.unesco.org/en/articles/guidance-generative-ai-education-and-research, accessed 18 September 2023

White, K, *It's Always Your Move: Purposeful progress for corporate career women* (Expert Author Publishing, 2018)

Online learning

Future Learn: www.futurelearn.com

Codecademy: https://codeacademy.com

Coursera: www.coursera.com

Pluralsight SKILLS: www.pluralsight.com

Useful websites

BCS for qualifications: www.bcs.org/qualifications-and-certifications

BCS Women: https://bcswomen.bcs.org

Business Analysts Info: www.bridging-the-gap.com

Black Girls Code: www.blackgirlscode.com

Cajigo: www.cajigo.com

Code First Girls: https://codefirstgirls.org.uk

Digital Eagles: www.barclays.co.uk/digital-confidence/digital-wings

Everywoman in Tech: www.everywoman.com

Institute of the Future of Work: www.ifow.org/resources/publications

Institute of the Future of Work: www.ifow.org/news-articles/time-uk-algorithmic-accountability-act

IT Infrastructure Library: www.coursera.org/articles/what-is-itil

Law Society: A guide to race and ethnicity terminology and language, www.lawsociety.org.uk/topics/ethnic-minority-lawyers/a-guide-to-race-and-ethnicity-terminology-and-language

National Careers Service: https://nationalcareers.service.gov.uk/job-profiles/business-analyst

Safe & The City App: www.safeandthecity.com

STEM Ambassadors: www.stem.org.uk/stem-ambassadors/find-a-stem-ambassador

StemConnext: https://stemconnext.co.uk

Stemettes: https://stemettes.org

Tech London Advocates: www.techlondonadvocates.org.uk

TechWomen100 Awards: https://wearetechwomen.com/techwomen100-awards

The Black Women in Tech: https://theblackwomenintech.com

The Skills Toolkit: https://nationalcareers.service.gov.uk/find-a-course/the-skills-toolkit

WeAreTechWomen: https://wearetechwomen.com

Women Who Code: www.womenwhocode.com

Acknowledgements

First, a big thank you to Hira Ali for believing in me. When I first shared my idea for this book, she liked the concept and encouraged me. Life has ups and downs, and things get in the way. Hira has been an incredible support during the writing and in the lead-up to the final stage of the book. She is a fantastic person, full of wisdom, who I would like to thank sincerely.

A massive thank you to the women who gave their time for the interviews with me: Dr Anne-Marie Imafidon MBE, Kasia Wojciechowska, Akua Opong and Asia Sharif. Your input was so good to make things come alive for the readers.

I thank my professional contacts, mentors, women in tech communities, WeAreTech founder Vanessa Vallely for the annual Tech100 awards and the amazing WeAreTheCity community that I have actively participated in for many years. Your support has been invaluable.

A thank you to the broad community of girls and women that surround us all, for anyone who nominated me for awards, and especially to those who follow me on all my social media platforms and engage with my posts. Also, to all those who have listened at the various conferences, topical panel discussions, events and podcasts that I have been invited to speak at or attend, especially those who asked insightful questions and responded with appreciative comments at these events. Thank you to those who invited me to be a part of these events, too. You have all been inspiring. Supporting one another on our journeys enables us to go further, as we are much stronger together.

Finally, thank you to my close family and friends, who are always so near, loving and dear to me.

I wish my late father and late mother were here to read this book. However, I know their enduring love and confidence in my brother and me has been the difference that has seen us achieve all we have… and will continue to achieve.

My dear late father and mother, I love you dearly and miss you constantly.

The Author

Sonal Shah is a *Forbes*-featured multiple award winner. She has over two decades of corporate experience and is a vice president in banking and founder of She Chose Tech. She has navigated the corporate and technology landscape, often as a minority in the rooms she entered. Recognising the gender gap in technology, Sonal passionately advocates for gender parity and empowers women within and outside her day job, delivering DE&I-inspiring keynote talks and coaching to accelerate their careers and unlock their potential.

Her prestigious awards include *Brummell Magazine's* Inspirational Women 2023, WeAreTheCity Rising Stars in Diversity and WeAreTech Tech100 for her work to increase the percentage of females in tech, inspiring and helping them join the industry and winner of the Women in Cloud empowHERaccess award for Community Leadership and Global Woman of Inspiration in 2022. Sonal was named in *Computer Weekly*'s Most Influential Women in 2022 and 2023.

She is a compelling speaker who frequently talks at industry-wide events for International Women's Day, careers panels and women in tech, among others. Sonal has been featured in *Forbes* and provides media commentary. Do feel free to keep in touch with Sonal via:

✉ Info@shechosetech.com

🌐 https://shechosetech.com